Translations of Responsibility

In 2020, a group of European researchers got a European Union (EU) grant to do a project called TRANSFORM. The objective of TRANSFORM was to integrate the principle of responsible research and innovation (RRI) into the research and innovation policies of three European regions: Lombardy, Brussels, and Catalonia.

This book tells the story of how TRANSFORM translated RRI into practice, all the way from philosophy of technology to EU policy jargon, to the project contract, and finally into the real-life events in these regions. Responsibility was translated in creative ways, with surprising goals and ambiguous outcomes. Armed with these stories, the book analyses the broader context of the desire for better governance of technoscience and draws two lessons: Firstly, that there is more governance than one may see at first sight, and secondly, that there is a need to rethink the borders of technoscience and the spaces in which it resides.

The book proposes to think of governance in technoscience, rather than governance of technoscience.

Thomas Völker is a postdoctoral researcher at the Centre for the Study of the Sciences and Humanities (SVT) of the University of Bergen, Norway. His research interests include practices of knowledge production and circulation in environmental governance and modes of experimenting with participatory democracy in policymaking.

Rasmus Slaattelid is Director and a professor at the Centre for the Study of the Sciences and Humanities (SVT), University of Bergen, Norway. His research focuses on translations between fields of knowledge, in the form of metaphors, models, and analogies.

Roger Strand is a professor at the Centre for the Study of the Sciences and Humanities (SVT) of the University of Bergen, Norway, and Co-Director of the European Centre for Governance in Complexity. His research focuses on uncertainty and complexity at the interfaces between knowledge and action.

History and Philosophy of Technoscience

Series Editor: Alfred Nordmann

Translations of Responsibility

Innovation Governance in Three
European Regions

**Thomas Völker, Rasmus Slaattelid,
and Roger Strand**

Routledge
Taylor & Francis Group
NEW YORK AND LONDON

First published 2024
by Routledge
605 Third Avenue, New York, NY 10158

and by Routledge
4 Park Square, Milton Park, Abingdon, Oxon, OX14 4RN

Routledge is an imprint of the Taylor & Francis Group, an informa business

Funded by Universitetet i Bergen.

ISBN: 978-1-032-44251-8 (hbk)
ISBN: 978-1-032-44252-5 (pbk)
ISBN: 978-1-003-37122-9 (ebk)

DOI: 10.4324/9781003371229

Typeset in Times New Roman
by KnowledgeWorks Global Ltd.

Contents

vi *Contents*

Preface – What is territorial RRI?

This book is the outcome of a three-year journey that was the project TRANS-FORM. The spelled-out version of this acronym is "Territories as Responsive and Accountable Networks of S3 through new Forms of Open and Responsible decision-Making." The project was funded by the Science with and for Society (SwafS) funding stream within Horizon 2020 and set out to "co-create more responsible approaches to innovation" in three European regions or territories.

Our responsibility in this project was to design and conduct the "monitoring and evaluation" process of the project. As this was a project on responsible research and innovation (RRI) it was clear to us from the start that a "traditional" monitoring and evaluation approach would not be able to do justice to the work done in the different regions. We had to do better, which for us meant – to use Donna Haraway's expression – "staying with the trouble" and to follow the work of our colleagues as closely as possible.

At this point, some readers might ask themselves why they should read another book on RRI. Admittedly, there are already a lot of those around. Now, we do not expect readers (at least not a lot of them) to be interested in reading approximately 200 pages of text about a project on RRI. Luckily, this is not what this book is about. It is about *more than that* – a phrase and idea that we will encounter at several points throughout the book. It is about unsold food and water quality in Brussels, waste management and endometriosis in Catalonia, and the quality of public administration in Lombardy. And, we argue, about the technoscience in all these things. It raises fundamental concerns about what Europeans think of as the emergence of European science and European modernity. It is also a book about the injustice and violence that arose together with European scientific, technological, and political orders, and about the attempts to somehow correct, manage, or govern what some have called a runaway train.

In this way, the book discusses RRI as a proposal to think about the relationship between society and technoscience, while also looking back on the history and politics of this responsibility as a policy "buzzword," to use Bernadette Bensaude-Vincent's phrase.

Crucially, this way of writing about RRI is also informed by the point in time at which we are writing it. RRI is dead (at least as an EU policy concept), and it is not yet clear what will come after. Thus, we are living in a moment in which there are no particular stakes, and the gloves can come off. It is a good time to take a step back and think about what has been done and what has been achieved with an overall share of 0.5 percent of the total Horizon 2020 budget that was devoted to the SwafS programme.

The book was written as a true collective endeavour, with three co-authors writing, reading, and re-writing its nine chapters. Moreover, it could only come to light because of the generosity, hospitality, and patience of our colleagues in the TRANSFORM consortium, led by Angela Simone and the Fondazione Giannino Bassetti in Milan, Italy. Our sincerest thanks go to all our TRANSFORM colleagues who so generously shared their experiences and opinions with us and who also endured our continuous probing into their practices which went beyond what a more "traditional" monitoring and evaluation work package would have demanded of them. This also meant that we politely – at least that was our aim – forced our colleagues to refine their expectations of what monitoring and evaluation means in the context of an RRI project. We are grateful that our colleagues from Lombardy, Catalonia, and Brussels allowed us to be part of their territorial journeys and were open enough to follow us on ours.

We are most grateful to Maria Skjelbred Meyer who helped us with getting the manuscript together in good style when help was most needed. Remaining errors and inelegant sentences are certainly our fault and not hers. Furthermore, we thank Alfred Nordmann and the editorial board for giving us the opportunity within this series to publish our work.

Writing a whole book is a cumbersome and costly affair. The empirical study presented in this book was, as already noted, part of the TRANSFORM project, which received funding from the European Union's Horizon 2020 research and innovation programme under grant agreement no. 872687. Some of the conceptual work, mainly contributions to Chapters 1, 3, and 9, by one of the authors (Roger Strand) was part of the SUPER MoRRI project, which received funding from the European Union's Horizon 2020 research and innovation programme under grant agreement no. 824671. The work, as well as the open access of this book, was as such funded by the European Union. We are obliged to inform the readers that the contents of this book are the sole responsibility of the authors and can in no way be taken to reflect the views of the European Commission. Indeed, upon reading the book, no reader is likely to believe that the contents in any way would reflect official EU views. Furthermore, the work presented in Chapter 8 was mainly funded by two project grants from the Research Council of Norway, namely the one for the Centre for Cancer Biomarkers (project 223250) and the Centre for Digital Life Norway (projects 248810 and 320911). Finally, the European Centre for Governance in Complexity is thanked for having funded the final lap of Maria Meyer's contributions when the university hiring bureaucracy proved unsurmountable.

Our empirical research in the TRANSFORM project was approved by the Norwegian Centre for Research Data, which was the Norwegian ombudsman for research data protection at the time.

Acronyms and abbreviations

ANT	actor-network theory
CS	citizen science
ELSI/ELSA	ethical, legal, and social/societal issues/aspects
EU	European Union
FGB	Fondazione Giannino Bassetti
RRI	Responsible Research and Innovation
S3	smart specialisation strategies
SSH	social sciences and humanities
SME	small and medium-sized enterprise
STS	science and technology studies
SwafS	Science-with-and-for-Society
TA	technology assessment
TRANSFORM	Territories as Responsive and Accountable Networks of S3 through new Forms of Open and Responsible decision-Making

1 The problem of responsibility in the governance of technoscience

Out to sea, facing the abyss

A book about "innovation governance in three European regions"? A book about the "governing of technoscience"? On what kind of planet, in what kind of society would anyone find such topics interesting and meaningful to the extent that they would like to read 200 pages about them? Why would someone spend a year writing 200 such pages?

The full answer to these questions is the book itself. The proof of the pudding is in the eating, and throughout the subsequent chapters, our readers are going to taste the rich details of some small-scale experiments at the interface between science and society. These experiments took place in Lombardy, Catalonia and in the Brussels Capital Region in the years 2020–2022 as part of a project funded by the European Union (EU), with the purpose of contributing to making research and innovation more responsible. To make sense of this purpose is, we believe, most of all a matter of providing such a richness of detail that the reader can understand and possibly empathise with the persons and organisations involved in the project. It is a matter of seeing how the project made sense to its actors and to us who wrote the book and who interpreted their endeavours in our own ways. The proof will be in the zooming in.

And yet, let us first search for sense by zooming out and by asking what kind of phenomenon the desire to "govern technoscience" is in the first place. What does it mean and from where does this desire come? Perhaps there could be a hundred different answers to these questions, taking a hundred different historical and philosophical perspectives. We shall tell the story in our way by aligning it to two well-known narrational strategies: We shall connect it to a *longue dureé* storyline around the idea of a scientific revolution and the emergence of modernity, and we shall play on the etymology of the word "govern," profiting from its nautical roots as a concept for steering a ship. Now that you have been warned, let us set out to sea.

The opening joke of Steven Shapin's (1998) book about the scientific revolution goes: "There was no such thing as the Scientific Revolution, and this is a

DOI: 10.4324/9781003371229-1

book about it." As with every good joke, its author had a serious intent. His point was that time travellers visiting 16th and 17th century Europe would have failed to encounter proponents and advocates of a scientific revolution. They thought of themselves neither as revolutionaries nor as scientists. The revolution is a post hoc historical construct that helps later generations make sense of the endeavours of Galileo, Descartes, Kepler, Newton, and the other heroes of science textbooks. There were such things as these endeavours, however, as well as their endeavourers. They described, depicted, and represented; in this sense there was science. They intervened, created, designed, and invented; in that sense there was technoscience. And in both cases, they searched for the new, they investigated and tested boundaries, going towards the boundaries and beyond them into outer astronomical spaces as well as inner natures of mechanical, optical, and mathematical objects. They explored.

And so these European men (and some women), who came to be celebrated as the heroes of later generations' European men (and possibly women), could be seen to take part in a broader culture, in an *Age*: The Age of Exploration. According to the narrative that we, the authors of this book who ourselves make up another such set of European men, were told and taught at school, it was an age in which Europe lit up from the Dark Ages. Europe was born again, we were told, released itself from the chains of medieval mediocracy and cast off to find new worlds. And new worlds were found, mapped, discovered, conquered, and colonised by explorers well-equipped with new tools for navigation and domination. To the extent that there was such a thing as the Age of Exploration, it was limitless, without boundaries. The farther away, the greater the glory. The more radical the approach, the larger the reward. For the sciences, the "New Organon," Bacon (1620) concluded *"ut vero ad interiora et remotiora Naturae penetretur, necesse est"*: It is necessary to penetrate the innermost and most distant of Nature's parts to be able to dominate and control her. For the conquerors on the sea, the fortunes could be made by killing those who owned them or selling those who were them.

Casting off and knowing no boundaries, one may come to face the abyss. In de Sousa Santos' (2007) analysis, the Europeans even created it. They (*we!*) developed and perfected what he called "abyssal thinking" by the construction of otherness – the Indians, the non-humans, *Nature* – for which our categories, norms, and boundaries do not apply or rather, for which it would be unthinkable to apply them. Hence, in the "colonies" we could plunder, rape, maim, kill, and destroy without – by our own standards – ever committing a crime. In the sciences, the penetration, domination and intervention into Nature could be justified by the practices of what Bruno Latour called the practice of purification, a distinct form of abyssal thinking whereby any political or moral qualm can be dismissed as ignorant. After all, the story went, science is only depicting, only discovering the truth that is already there. Science is not politics; nature is not culture.

In a book about governance, we hope for patience as we push the nautical metaphor. The conquistadores, of course, were literal sailors on wooden ships travelling aquatic oceans. The ship of our scientific heroes, or possibly their Armada, was more abstract. It was close to the essence of modernity and enlightenment itself: The New Organon, the truth machine necessary and sufficient for omnipotence, for "where the cause is not known, the effect cannot be produced" (Bacon 1620). The literal sea explorers had to navigate, not necessarily always to arrive at a specific place since finding new land indeed might be the purpose of the venture itself, but to sail safely through the dangers of deep and shallow waters. The nautical root to governance is tied to this idea of arriving safely, of steering clear of cliffs – the *riescos* – and storms, to protect oneself and the crew from drowning here and now. Of course, for those who were discovered, the navigational success immediately resulted in the opposite of safety. Through abyssal thinking, however, Europe could dissociate from that fact, as if to anticipate the Gestalt of Eichmann in Jerusalem, who in Hannah Arendt's interpretation was the personified banality of evil, the clerk who never asked questions, who stayed attentive to rules and orders but lost the ability to be mindful in the proper, moral, human sense. Through this figure, she could explain the horrors of the World Wars, whereby the violence so long cultivated in the peripheries overflowed and returned to make Europe itself collapse and sink into a moral and civilisational abyss. A five centuries long dureé is needed to see this shipwreck in its entirety.

Is science on a disastrous course? Is science shipwrecking and the crew mindless? Is anyone governing science? Again, there are the stories we were taught at school, stories that invoke a different idea of governance than the helmsman at the wheel. As Michael Polanyi (1962) would have us believe, the Republic of Science is self-governed, it is a perfect nobility or priesthood created and constituted by its pure and rigorous truthfulness. As long as *Homo depictor*, the man of science, stays clear of the cliffs of politics, money, and other corruptions, he will not shipwreck, and the Republic of Science will continue on its triumphant journey of exploration.

What Polanyi did not consider, however, is that the scientists are also *Homo faber*. What happens in their peripheries? What might not be seen in the blind zone of scientific abyssal thinking? Might science cause violence that overflows and returns on us? That question is one way of answering the question about why anyone would write 200 pages about innovation governance in three regions.

Thomas Kuhn (1962) opened his treatise on scientific revolutions with a remark that was anything but a joke. "History," he wrote, "if viewed as a repository for more than anecdote or chronology, could produce a decisive transformation in the image of science by which we are now possessed." The transformation in question was to realise that *Homo depictor*'s truth machine is operated by real human beings and not Laplacian intellects with a view from nowhere. Truth was truth, but it was conditional to the paradigm in which it was discovered and incommensurable with truths from other paradigms. The full implications of the Kuhnian exorcism have kept science studies busy ever since 1962. With or without ascribing to the dubious historiography of *Structure,* scholars described science ever more precisely as human practice and culture. With that in mind, there was no way to prevent the trains of thought that led to seeing that scientific knowledge creates social problems (Ravetz 1971), even if Kuhn personally did what he could to stop it (and even stop Ravetz himself; Fuller 2000). All it took to realise that science both represents and intervenes (Hacking 1983), and that neither activity is innocent and beyond moral appraisal, was to get over one's love affair with science. As soon as the spell was broken, one was set free to discuss if not science indeed is technoscience, how it connects to the industrial-military complex, how it deliberately is used to disrupt society as an engine in the acclaimed vehicle of "disruptive innovation," and how it played a causal role in the development of climate change, loss of biodiversity, pollution, soil erosion, and virtually all major environmental problems. If Europe suffered moral and civilisational collapse in World War I (WWI) and World War I (WWII), it was also because of the unprecedented technological efficiency in murder and violence that relied on chemistry and physics. As for the transformation of the world that happened with $E = mc^2$ and the creation/discovery of nuclear chain reactions, first patented in 1934 by Leo Szilard, nobody explained the significance better than Szilard himself:

> On MARCH 3, 1939, Dr. Walet [sic] Zinn and I, working on the seventh floor of the Pupin Building at Columbia University, completed a simple experiment to which we had been looking forward rather eagerly. Everything was ready, and all we had to do was to lean back, turn a switch, and watch the screen of a television tube. If flashes of light appeared on the screen, it would mean that neutrons were emitted in the fission of uranium, and that in turn would mean that the liberation of atomic energy was possible in our lifetime. We turned the switch, we saw the flashes, we watched them for about ten minutes – and then we switched everything off and went home. That night I knew that the world was headed for sorrow.
>
> (Szilard 1945)

The world was headed for sorrow but somehow that fact made only a moderate impression on those who knew and who continued to head forwards. Abyssal thinking may have different natures; there is a difference between Eichmann's banality and Szilard's melancholy. Sir Isaac Newton apparently thought of the sextant but did not promote the invention. It seems farfetched, however, to speculate that he thought of its use in navigation and colonisation and knew that the world was headed for sorrow. Szilard knew and stayed with his moral troubles

(Strand 2010). A generation later, the acclaimed scientist and inventor Marvin Minsky replaced sorrow with something more nihilist:

> I find it appalling how many people are willing to accept the bad deal they have been given. We ought to be more insistent about improving our brains and our bodies. [...] I find it even more annoying that we have to live only a hundred years just because of a few evolutionary mistakes. When we design new forms for ourselves, we will describe our intentions along with the plans.
>
> *QUESTION FROM THE AUDIENCE:* Do you anticipate the development of a hacker culture of nanotechnology?
>
> *MINSKY:* There are hackers, and there are crackers. [...] It seems to me that a way must be found to keep things open enough so that we can catch malicious people before they can do anything too bad. Accomplishing that will not be easy. We might have to give up our privacy. There are terrible things in the universe. Quasars, for example, appear to be galaxies that exploded because something bad happened there. I wonder how many of those were science-fair projects that got out of hand.
>
> (Minsky, in Krummenacker and Lewis 1995, 195)

Our technoscience may blow up the galaxy but so be it. It does not matter as long as we are heading forward, forward, out on the sea, into the abyss. The tale told in this way, technoscience à la Minsky is not banal evil, it is raging madness. Its essence is not that of a killing spree like those of the conquistadores, it is a collective suicide mission for the fun of it, a mission driven by boundless curiosity which is not satisfied until it kills the cat and everybody else. Nobody is at the wheel and nobody should be at the wheel. The Republic of Technoscience is organised irresponsibility and accelerated recklessness, a ship with no captain, a run-away train.

Enter the desire to govern technoscience and to make research and innovation responsible and not reckless. This story is rarely told in such dramatic tone and with so sweeping claims within the world of policy and governance institutions. In the world of art, literature, and movies, the drama has been rehearsed for a long time, at least since Mary Shelley wrote her novel about Doctor Frankenstein and let her eponymous hero explain the moral of the story, namely that a wise man would never "allow passion or a transitory desire to disturb his tranquillity. I do not think that the pursuit of knowledge is an exception to this rule" (Shelley 1818). This motif has existed in culture at least since the myth of the Golem and is still a main ingredient in successful science fiction. In political and public debate within modern society, however, the motif is offensive. It debases the love of the many lovers of science, such as the thousands of members of social media publics under the title "I Fucking Love Science" (García Casañas 2021). It offends our scientific, industrial, and political elites. This book speaks of RRI, of responsible research and innovation, a concept and a policy principle that was born in and around the European Commission in the late 2000s and early 2010s. A more focused history of RRI is presented in Chapter 3. What is striking, however, is how polite and cautious the introduction of RRI was. Whenever those who f***ing loved science were sceptical against ethics or responsibility or any other attempt at governing technoscience, the ethicists, philosophers, and policymakers made sure to explain that there was no implicit accusation being made anywhere. To call for ethics was not to insinuate insufficient moral standards. To argue for RRI was not to call out irresponsibility and recklessness. No, it was to improve our institutions so that already quite favourable states of affair would become even more favourable. Science would become even better aligned with society, and the public would love and participate in science even more than the Eurobarometer had shown them to do, with the embarrassing exception of some biotechnologies. For anyone who was given such statements and proclamations without prior exposure to critiques of modernity and critiques of science, it would have to have been infinitely difficult to find anything of interest in them and figure out that there was anything real at stake. It would be like guessing the taste of coffee from drinking latte macchiato. "Governance of science" would sound as any other meaningless Euro-speak policy jargon with words invented by consultants as the key ingredient of their business model.

Captain or body politic?

In philosophical terms, the Crisis of the European Sciences was analysed well before Szilard switched off the lights and went home. Husserl and Heidegger had the diagnosis ready before WWII, and it was elaborated further after the war by the Frankfurter School as well as other philosophers and sociologists who admittedly shared the subject position of the abyssal thinkers – European white males with comfortable living standards. One approach to the question of governance that accordingly emerged at least in academia from the 1960s and onward was to abandon the ship of European science altogether or at least give it a complete makeover. From other standpoints – feminist, queer, Marxist, religious (in particular Islamic), post-colonial, ecological – it might be possible to build a new ship, new bodies, cultures and practices of knowledge that did not destroy the world. These intellectual developments are at least as important as the phenomena that we discuss in this book, and there are also connections between them. There has been some uptake. Our focus, however, remains on Europe and its own intellectual and institutional struggles to come to grips with the insight that the science and technology that it loves so much, can also be harmful.

There is rich philosophical and historical literature that in part describes, in part participates in, these struggles. The contemporary version of it is discussed in Chapter 3. The purpose of this introductory chapter is merely to convey a sense of why the problems we are presenting are important, interesting, and hard. For this purpose, we need to simplify. One brilliant simplification was provided by one of the protagonists, if not the main one, in the debates over what came to be known as RRI, namely the philosopher of technology René von Schomberg. From 1998 to 2022, von Schomberg worked as a civil servant for the European Commission's Directorate-General for Research and Innovation (DG RTD, formerly known as the Directorate-General for Research, Technology, and Development). We shall return to his intellectual and political achievements in Chapter 3. His "Vision of Responsible Research and Innovation" (Schomberg 2013), however, touches on a central theme, namely the need for collective and institutional approaches to the problem of governance of technoscience. In this book chapter, von Schomberg tells the story of the *Passarola,* an invention made (or at least designed) by a Portuguese priest called Bartolomeu Lourenço de Gusmão at the beginning of the 18th century. Passarola apparently means "ugly bird" and the invention was a sort of airship. Bartolomeu presented the idea to the Portuguese king John V, making clear that the airship could create a lot of opportunities but also risks, notably that it might allow perpetrators of crimes to easily flee the country and thereby escape law enforcement. Von Schomberg tells how John V gave Bartolomeu the exclusive right to develop the airship further and – and this is the crux of the story – installed the death penalty to anyone who tried to copy the invention.

Von Schomberg discusses the king's solution to governing the invention as a governance regime that could work in a quite simple context of a solitary inventor and a monopoly of power. Responsible governance of technology could then be a matter of prohibition, monopoly, and hard law. The technology could be prohibited or alternatively be allowed only for those with a license to use it responsibly. This governance regime still exists for some technological fields that are particularly lethal, such as nuclear technology and more generally what is demarcated into the military domain. There are at least two reasons why it is insufficient for responsible governance of technoscience in modern societies, however. Firstly, von Schomberg rightly pointed out that most modern innovation systems are too vast and too tightly coupled to markets to allow for sovereign control. Indeed, the economic systems are based on principles of free enterprise and that innovation and creativity should be as free from intervention as possible, as long as they satisfy basic requirements of efficacy, quality, and safety. More than often, if a national government contemplates a strict regulation against a technology, it is met with the counterargument that it will be developed somewhere else anyway.

Secondly, even if one somehow manages to resist the pressure from the market economy, it is difficult to justify hard regulations. How can the king know that it is the right decision to prohibit the technology? In the scholarly debates on the possibility of Islamic science, one line of argument was that scientific claims and technological developments could be checked for their compliance or coherence with the Quran. If they could be seen to align with the words of the Quran, they would be *halal.* If not, they would be *haram* and should be regulated against. Outside that particular religious worldview, however, this solution would not only be unfeasible. It would be irrational and ridiculous. This observation connects to the depth of Polanyi's notion of the Republic of Science. Polanyi was not arguing for scientific privilege. He was arguing against the notion of a Captain who mysteriously knows, or rather a dictatorship that trumps rational and truthful discourse, which for him was scientific discourse. If society is led as a ship where one individual, one party, or one other unaccountable source of authority gets to be the helmsman, this indeed leads towards the abyss – the moral catastrophe of the Moscow trials or the scientific catastrophe of Lysenko's ban on Darwinian evolutionary theory.

This is why there should be a Republic of Science that self-governs: It is a means against a sort of abyssal thinking that sacrifices truthfulness and the idea of truth itself. And the elitism that was implicit in the seemingly egalitarian and universalist conception of science found not only in Polanyi but also his contemporaries Robert K. Merton and Karl Popper, if not in logical positivism itself, was an attempt at defending and demarcating pursuits of truth and truthfulness from violent authoritarian forces. This is what they could see; somehow, they were not able to integrate that already the non-Nazi, non-Soviet, European science was in crisis. They were on a different mission.

So, both in terms of power and knowledge, attempts at improving the governance of technoscience had to relate to the Republic of Science. This was much more than a matter of Polanyi or Merton; it was a matter of which discourse that dominated in the Western world and beyond. Even in the Soviet Union, Lysenko finally came to his demise. The solution to be sought was to improve and refine the Republic of Science by making its citizens more aware of their own blind zones and more responsible, and perhaps by democratising it in the sense of increasing its contact with the larger society and the real body politic. President Dwight D. Eisenhower anticipated the challenge perfectly in his Farewell Address in 1961, more than half a century before the notion of RRI emerged and almost 40 years before Gibbons (1999) described the change in the social contract of science. The clarity in Eisenhower's formulation is remarkable and came right after the more famous part of the speech, where he warned against the power of the industrial-military complex (see also Funtowicz and Strand 2011):

Akin to, and largely responsible for the sweeping changes in our industrial-military posture, has been the technological revolution during recent decades. In this revolution, research has become central; it also becomes more formalized, complex, and costly. A steadily increasing share is conducted for, by, or at the direction of, the Federal government.

Today, the solitary inventor, tinkering in his shop, has been overshadowed by task forces of scientists in laboratories and testing fields. In the same fashion, the free university, historically the fountainhead of free ideas and scientific discovery, has experienced a revolution in the conduct of research. Partly because of the huge costs involved, a government contract becomes virtually a substitute for intellectual curiosity. For every old blackboard there are now hundreds of new electronic computers. The prospect of domination of the nation's scholars by Federal employment, project allocations, and the power of money is ever present and is gravely to be regarded.

Yet, in holding scientific research and discovery in respect, as we should, we must also be alert to the equal and opposite danger that public policy could itself become the captive of a scientific-technological elite.

It is the task of statesmanship to mold, to balance, and to integrate these and other forces, new and old, within the principles of our democratic system-ever aiming toward the supreme goals of our free society.

(Eisenhower 1961)

When society becomes a knowledge society, are not all citizens also scientific citizens? Could there be universal suffrage in the knowledge society? Could politicians and laypersons meaningfully engage in the ethics and politics of science, through ethics boards, technology assessments, science cafés, and the broad range of practices called public participation and public engagement with science? Such ideas were gradually explored in the late 1960s and into the 1970s and 1980s. At the same time, research funding policy became more actively governing, having noted that the "linear model" advocated by Vannevar Bush and his generation was not a fair representation of reality. Science did not give most value for money if the money was pumped into basic research and the funding agencies operated in the style of "fund and forget."

It would be interesting to know if it ever happened that universal suffrage was welcomed by those who lost their privilege. In the case of the Republic of Science, the battles between those who want to govern science and those who f***ing love it are still ongoing. Rommetveit (2007) recalls how even bioethics was met with resistance in the US. A telling example is Senator Walter Mondale's efforts to establish a Presidential study commission on "organ transplantation, genetic engineering, behaviour control, experimentation on humans, and the financing of research" (Jonsen 1998, 91). Mondale's first initiative in 1968 failed because of opposition from scientists and medical doctors. His second attempt in 1971 came out successful, however not without resistance:

... all we are proposing here is to create a measly little study commission to look at some very profound issues...I sense an almost psychopathic objection to the public process, a fear that if the public gets involved, it is going to be anti-science, hostile and unsupportive.

(quoted from Jonsen 1998, 94)

Meanwhile, since the 1970s, the world of science became ever more technoscience with an ever stronger interest in engineering and technology. Public funding and public interest directed itself ever more towards sciences that become technologies, such as biotechnology, nanotechnology, and information and communication technologies. Furthermore, the Frankenstein character of technoscience went from being mere science fiction to reality, with cloning, research on human embryos, genetic modification, brain-machine interfaces, etc. In the United States, the nanotechnology funding discourse became explicitly transhumanist, speaking of technological enhancement of the human species in physical, cognitive, and even moral terms (Roco and Bainbridge 2003). The horror this was met with in the EU resulted in a brave countermove, namely a vision for governing the research trajectories of emerging technologies with European values and needs at the steering wheel (European Commission 2005).

European governance and the structure of this book

The concept of governance as tied to the body politic rather than the helmsman is relatively recent. On the political scene, it came to prominence in the 1990s with the Carlsson Commission and its report on global governance (The Commission on Global Governance 1995). In political science, the term had already been used for some time as an analytical concept.

The European Commission published its White Paper on European Governance in 2001. It laid out five principles for good governance: Openness, participation, accountability, effectiveness, and coherence. The style of this document was remarkable in that it took candid critical look at the relationship between EU institutions and citizens. Indeed, the opening statement read:

Today, political leaders throughout Europe are facing a real paradox. On the one hand, Europeans want them to find solutions to the major problems confronting our societies. On the other hand, people increasingly distrust institutions and politics or are simply not interested in them.

The white paper went on to argue that democracy has to be revitalised by a more open and less top-down hierarchical attitude from the political institutions and by wider and more inclusive practices of public participation "throughout the policy chain" (European Commission 2001, 8). For policy principles such as RRI, the white paper served both as a source of inspiration and a high-level policy reference.

At the same time, it would not be historically correct to claim that the push for good governance in the EU mainly was due to critiques of and qualms with technoscience. Biotechnology and the public controversies around the use of genetically modified organisms (GMOs) in agriculture were part of the political context. These controversies led to the so-called "de facto moratorium" on GMOs in the EU from 1998 to around 2004 (Lieberman and Grey 2006). Furthermore, there was the scandal of the transmission of mad cow disease into humans through infected beef, which mainly took place in the United Kingdom and which triggered top-down actions to "restore" trust in science by means of public engagement exercises (Wynne 2006). Science was not exempt when the white paper boldly stated that people increasingly distrust institutions. Still, the main concern of the white paper was the increasing troubles that the European project, that is, the formation of the EU as a supranational political union ran into. These troubles consisted above all in the lack of citizen support. Participation in European elections was disappointing. Much worse, however, was the emergence of so-called wrong answers, in particular in the referendums about the Maastricht Treaty. The scandals addressed by the white paper were not so much the GMOs and mad cows as the two referendums in 1992 that resulted in the Danish "no to Maastricht" and the French "petit ouí" with only a 51.05% majority of the French voters (Lewis-Beck and Morey 2007).

In sum, looking backwards from 2010, there are several historical scales that can allow us to think of RRI as born out of this mess co-produced by as well as co-producing science, technology, and society. The short timescale included worries about (legitimate or illegitimate) public concerns around the projects of scientific and political elites, including the prestigious technoscientific endeavours such as biotechnology and nanotechnology, EU research and innovation policies, and all the way up to the European project itself. On the mid-range timescale, one can tell a story whereby the world was almost blown up by advances in physics in the middle of the 20th century, and where the series of technologically created or accelerated crises seemed to grow longer by the day. And finally, as we have tried to do in this chapter, there is the longue dureé perspective of intellectuals from Husserl to post-colonial thinkers who considered the European scientific and technoscientific project to be reckless and irresponsible, indeed abyssal, from the very beginning.

When faced with a mess, it is not uncommon to try to do something about it – change something, clean up something. The whole point with creating the RRI principle was to encourage action. Von Schomberg (2013) developed his ideas of how to create collective responsibility within the research and innovation system, that is, reforming the Republic of Science into responsible self-governance by some mild policymaking from above. Jack Stilgoe, Richard Owen, and Phil Macnaghten (2013) developed a British alternative, based on the sub-principles of anticipation, reflexivity, engagement, and responsiveness. As we shall dive deeply into later in this book, all of these philosophical innovations got too abstract to be taken up by the administrative machine of DG RTD, which instead devised a set of so-called RRI "keys" – ethics, public engagement, science education, open access, gender equality and, most inter-estingly, a key simply called governance. What they were supposed to entail, was also a rather open question.

This mess, the perceived need to act on it, and the challenges of doing so, is what this book is about. It is about studying how RRI is *translated* into practice, even if its proponents speak of "implementation" and then constantly get disappointed. What we mean by translation is the topic of Chapter 2. It looks closer at the origin of RRI as something to fund European projects on, in Chapter 3. It zooms in at a particular EU project, a Horizon 2020-funded project called TRANSFORM, that operated in three European regions, namely Lombardy, Catalonia, and the Brussels-Capital Region. We zoom in on each of these three regions in Chapters 4, 5, and 6, respectively. Chapter 7 is the longest one, as it digests and analyses the TRANSFORM experiences. As it turns out that TRANSFORM was a project that translated governance of technoscience in ways that made it hard to see the presence of science as well as technosci-ence, we contrast the TRANSFORM experience with two other RRI projects in which the hi-tech was more visible. Chapter 8 is devoted to this contrast. Finally, Chapter 9 asks what we have learnt from these experiences about the governance prospects. It suggests a way out of the abyssal thinking by seeing technoscience everywhere and proposing a shift from governance of technosci-ence to **governance in technoscience**.

References

Bacon, Francis. 1620/1994. *Novum Organum; With Other Parts of the Great Instaura-tion*. Chicago: Open Court Publishing.

Eisenhower, Dwight D. January 17, 1961. "Farewell Address by President Dwight D. Eisenhower." Final TV Talk 1/17/61 (1), Box 38, Speech Series, Papers of Dwight D. Eisenhower as President, 1953-61, Eisenhower Library; National Archives and Records Administration. https://www.archives.gov/milestone-documents/president-dwight-d-eisenhowers-farewell-address.

European Commission. 2001. "European Governance: A White Paper." COM(2001)428. https://ec.europa.eu/commission/presscorner/detail/en/DOC_01_10.

_____. 2005. Directorate-General for Research and Innovation, Alfred Nordmann. "Converging Technologies: Shaping the Future of European Societies." Publica-tions Office. https://op.europa.eu/en/publication-detail/-/publication/7d942de2-5d57-425d-93df-fd40c682d5b5.

Fuller, Steve. 2000. *Thomas Kuhn: A Philosophical History for Our Times*. Chicago: University of Chicago Press.

Funtowicz, Silvio, and Roger Strand. 2011. "Change and Commitment: Beyond Risk and Responsibility." *Journal of Risk Research* 14 (8): 995–1003.

García Casañas, Cristina. 2021. "Don't They Understand Climate Science? Reflections in Times of Crisis in Science and Politics." *Public Understanding of Science* 30 (8): 947–61. https://doi.org/10.1177/09636625211011882.

Gibbons, Michael. 1999. "Science's New Social Contract With Society." *Nature* 402: C81–C84. https://doi.org/10.1038/35011576.

Hacking, Ian. 1983. *Representing and Intervening: Introductory Topics in the Philosophy of Natural Science*. Cambridge: Cambridge University Press.

Jonsen, Albert R. 1998. *The Birth of Bioethics*. Oxford: Oxford University Press.

Krummenacker, Markus, and James Lewis. 1995. *Prospects in Nanotechnology: Toward Molecular Manufacturing*. New York: John Wiley & Sons.

Kuhn, Thomas S. 1962. *The Structure of Scientific Revolutions*. Chicago: The University of Chicago Press.

Lewis-Beck, Michael S., and Daniel S. Morey. 2007. "The French 'Petit Oui': The Maastricht Treaty and the French Voting Agenda." *The Journal of Interdisciplinary History* 38 (1): 65–87. https://doi.org/10.1162/jinh.2007.38.1.65.

Lieberman, Sarah, and Tim Grey. 2006. "The so-Called 'moratorium' on the Licensing of New Genetically Modified (GM) Products by the European Union 1998–2004: A Study in Ambiguity." *Environmental Politics* 15 (4): 592–609. https://doi.org/10.1080/09644010600785218.

Polanyi, Michael. 1962. "The Republic of Science: Its Political and Economic Theory" *Minerva* 1 (1): 54–73.

Ravetz, Jerome R. 1971. *Scientific Knowledge and Its Social Problems*. Piscataway: Transactions Publishers.

Roco, Mihail C., and William Sims Bainbridge. 2003. "Overview: Converging Technologies for Improving Human Performance". In *Converging Technologies for Improving Human Performance: Nanotechnology, Biotechnology, Information Technology and Cognitive Science*, edited by Mihail C. Roco and William Sims Bainbridge. Springer. https://doi.org/10.1007/978-94-017-0359-8.

Rommetveit, Kjetil. 2007. *Biotechnology: Action and Choice in Second Modernity*. Bergen: University of Bergen.

Schomberg, René von. 2013 "A Vision of Responsible Research and Innovation." In *Responsible Innovation: Managing the Responsible Emergence of Science and Innovation in Society*, edited by Richard Owen, John Bessant and Maggy Heintz, 51–74. John Wiley & Sons. https://doi.org/10.1002/9781118551424.ch3.

Shapin, Steven. 1998. *The Scientific Revolution*. Chicago: The University of Chicago Press.

Shelley, Mary. 1818/2006. *Frankenstein*. London: Penguin.

Sousa Santos, Boaventura de. 2007. "Beyond Abyssal Thinking: From Global Lines to Ecologies of Knowledges." *Canadian Parliamentary Review* 30 (1): 45–89.

Stilgoe, Jack, Richard Owen, and Phil Macnaghten. 2013. "Developing a Framework for Responsible Innovation." *Research Policy* 42: 1568–80. https://doi.org/10.1016/j.respol.2013.05.008.

Strand, Roger. 2010. "Leo Szilard: Immoral Science - Moral Fiction?" In *The Art of Discovery*, edited by Margareth Hagen, Randi Koppen and Margery Vibe Skagen, 171–80. Aarhus: Aarhus University Press.

Szilard, Leo. 1945. "We Turned the Switch." *The Nation* 156 (December 22): 718–19.

The Commission on Global Governance. 1995. *Our Global Neighbourhood: The Report of the Commission on Global Governance*. Oxford: Oxford University Press.

Wynne, Brian. 2006. "Public Engagement as a Means of Restoring Public Trust in Science–Hitting the Notes, but Missing the Music?" *Public Health Genomics* 9 (3): 211–20. https://doi.org/10.1159/000092659.

2 Territorial RRI as translation

From implementation to translation

In the previous chapter, we gave a brief introduction to the general idea of Responsible Research and Innovation (RRI) as it emerged in European policy-making. Before we can tell our story about the work in the different regional TRANSFORM clusters, it is necessary to develop our conceptual vocabulary.

In this chapter, we develop a lens for analysing territorial RRI projects that aims to shift the discussion away from questions of *implementation* to questions of *translation* as the term is used in the field of science and technology studies (STS). In our experience of working in various RRI projects, RRI was never simply *implemented* according to plan, and with the anticipated results. One possible explanation for this is that RRI originated as a policy concept and is built on ideas about the transformation of the research and innovation (R&I) system at a level of abstraction and idealisation that does not allow for straightforward implementation in concrete cases, any more than say, virtue ethics can be implemented in real estate markets. An empirical indication of this implementation challenge is the plethora of academic papers from the fields of STS and applied ethics that detail how exactly things did not go according to the RRI plan (Åm 2019).

However, it is important to remember that even if things do not go according to plan, still things happen. A philosophical idea can be *misunderstood* in the process of being translated into a workable operationalisation. A set of *RRI keys*, with their origins in what might be called an administrative coincidence in the organisation chart of the European Commission, might in some cases take on a new meaning and inspire the work of local innovators in surprising ways. The puzzlement and possibly frustration amongst a group of university researchers about the *real meaning* of RRI may lead to a fruitful debate about notions of anticipation and reflexivity. The list could be extended. This book is about how both hermeneutic processes of interpretation and institutional operationalisation contribute to shaping different situated translations of RRI, and in doing so sometimes perhaps even achieve transformative effects.

DOI: 10.4324/9781003371229-2

We use the term translation to think about what is commonly referred to as *implementation*. The central issue with implementation as both a concept and process is that it is premised on the idea that there actually is a right way of doing RRI; an essence of sorts or a core set of RRI principles if you will. Such principles may then be followed the right or the wrong way, which means that one can succeed or fail in attempts of implementing RRI. However, most scholars, practitioners, and policy officers tend to agree that there is no single valid definition of RRI. Despite many attempts of defining it, the concept remains a moving target. There are different RRI keys emphasised by the European Commission (2015) and there are frameworks focusing on principles like anticipation, reflexivity, inclusion and responsiveness (Fitjar, Benneworth, and Asheim 2019), but none of the elements in these frames are exclusive to RRI, nor do RRI projects apply all of the keys or principles. Therefore, as we will argue below, it is more productive to provide thick descriptions of processes of translation, than to assess the success of *implementing* RRI.

Translation 101 – literary and sociological perspectives

When we speak about translation in everyday language, we usually refer to the act of converting a word or a text from one language into another. Understood in this way, translation is "a technology of literary replication that engineers textual afterlife without recourse to a genetic origin." (Apter 2006 cited in Barry 2013). To translate between languages implies an ambition to preserve as much as possible of the original *meaning*, while at the same time acknowledging that in the process new meaning is *added* and that some of the original meaning is *lost*. Translation, therefore, must be thought of as a dual process of preservation of meaning on the one hand and modification on the other.

The term, however, is used also to describe non-linguistic processes of transfer. It is an important concept in actor-network theory (ANT), developed amongst others by Michel Callon, Bruno Latour, and Steve Woolgar. Building on work by Michel Serres and Michel Callon, Bruno Latour uses translation to describe processes of "displacement, drift, invention, mediation, the creation of a link that did not exist before and that to some degree modifies two elements or agents" (Latour 1994). The way, for instance, in which the relationship between a human and a tool creates a link that may modify the purposes and goals of both the tool and the human. In the context of ANT translation thus involves modification and exercise of power, stabilisation of networks, which always means the stabilisation of one version of the world and not others.

More recently the idea of translation has been used productively in the assessment of engagement activities. The focus on translation here allows for tracing how standardised methods or tools for engagement *travel* (Soneryd 2015; Soneryd and Amelung 2016; Laurent 2017; Konopásek, Soneryd, and Svačina 2018). We argue that it can also be a useful conceptual lens to think about territorial RRI practices.

Before we get there, however, we need to delve a bit deeper into the different ways translation is understood and used in STS. The term is used to describe the activities of researchers and scientists during the process of substantiating certain knowledge claims. The interesting point here is that these activities are seen to involve quite a bit more than what is usually talked about with regard to scientific *findings* or *discoveries*. This is why authors working in STS are talking about a "sociology of translation" (Callon 1986). While terms such as *findings* and *discoveries* indicate a rather passive role of scientists and researchers when nature is revealing some underlying truths to them, proponents of a sociology of translation would argue that to establish something that is accepted as *truth* involves building a network of human and non-human actors, that is, translation. Importantly, such a translation always implies establishing a position from which to legitimately speak for others. What is later called "obligatory passage point" is already carved out quite clearly in an early description of translation from 1981:

> By translation we understand all the negotiations, intrigues, calculations, acts of persuasion and violence thanks to which an actor or force takes, or causes to be conferred on itself, authority to speak or act on behalf of another actor or force. 'Our interests are the same', 'do what I want', 'cannot succeed without going through me'. Whenever an actor speaks of 'us', s/he is translating other actors into a single will, of which s/he becomes spirit and spokesman. S/he begins to act for several, no lonqer for one alone. S/he becomes stronger. S/he grows.
>
> (Callon and Latour 1981, 279)

This position is further developed in one of the most influential texts about translation in this literature, which is almost 40 years old yet still as relevant as when it was published. In his study about scallops and fishermen in the French St. Brieuc Bay, sociologist Michel Callon uses the term to describe a controversy about the reasons for the decline of the scallop population in St. Brieuc Bay (Callon 1986). In doing so he develops a material-semiotic sociology of translation.

Now what does this mean? When Callon talks about translation, he is talking about processes through which different actors try to find allies for building networks. Such allies – they are also referred to as "actants" – help stabilise certain knowledge claims, and thus certain accounts of the world. Callon is interested in the relation between knowledge and social order, or put differently, in the role of "science and technology in structuring power relationships."

This means tracing "the simultaneous production of knowledge and the construction of a network of relationships in which social and natural entities mutually control who they are and what they want." (Callon 1986, 203) Typically for this kind of ANT, this notion of translation also works as a counterpart to a conception of *truth* as grounded in a relationship of correspondence between a statement and a fact, or state of affairs, stressing that it is rather through a series of translations that knowledge claims gain power and stability and become *truths* (see also Latour 1999), not through a privileged relationship of correspondence to independent facts.

Callon distinguishes what he calls "four moments" of translation: These are problematisation, interessement, enrolment, and mobilisation. Problematisation is the definition of relevant human and non-human actors and their relations in relation to a certain problem. In the case of the scallops at St. Brieuc Bay, this problem was how to increase the production of scallops by new ways of cultivating. Interessement refers to the process in which the identities and relations between the different actors are stabilised together with their shared goals, or in Callon's words, when "the allies are locked into place" (Callon 1986, 206). In the next step, the roles of the different actants need to be defined and coordinated. This moment of the translation is called "enrolment" and simply refers to the multiple negotiations going on so that the different actors become allies and act according to plan. Finally, mobilisation means that some of the actors – in Callon's study the scientists studying the scallops – become spokespersons who represent others and are granted the authority to speak in their name.

What is translated here then is not a text but the identities, goals, and problem framings of a particular set of actors. Callon uses the term to describe the actions of a group of scientists to establish a certain version of a problem. What is important to note here is that such a translation is never fully stabilised, there is always the risk that something does not go according to plan. New controversies may come up and to re-negotiate the different roles may become necessary. Translation thus always bears the risk of betrayal.

This point is developed later by John Law, who defines translation as the combination of *traduction* and *trahison*, that is, the elements of similarity and betrayal in each act of translation (Law 2003). He applies this understanding to a description of different versions of ANT and thus argues against the possibility of a true or faithful translation. Law points out that translation is always to some extent a betrayal; it always contains within itself similarity *and* difference. This point is crucial when thinking about implementation as translation.

The twist – or conceptual novelty – is the principle of symmetry that Callon applied in this description. He writes about scallops in the same way he writes about scientists or fishermen. A certain reality, or more precisely a certain distinction between nature and society, for him is the outcome of the process and not the starting point.

In a similar way but focused more on questions of social order, Bruno Latour, in his study about hotel keys and the development of the Kodak camera, uses the notion of translation to describe networks of actants and how they shape a program of action (Latour 1991). *Translation* then refers to changes in the program of action (and their related networks), that is, when a weight is added to a hotel key or when the hotel manager explains to guests why they should leave the keys in the hotel lobby. Translation here is about the creation of new links between different actants that in turn also transform these entities. Following from that, such a focus on different translations becomes a methodological approach:

> If we display a socio-technical network – defining trajectories by actants' association and substitution, defining actants by all the trajectories in which they enter, by following translations and, finally, by varying the observer's point of view – we have no need to look for any additional causes.
>
> (Latour 1991, 129)

The importance of varying the point of view that Latour mentions here has been highlighted already earlier by Star and Griesemer in their study of Berkeley's Museum of Vertebrate Zoology (Star and Griesemer 1989). In this paper, they build on Callon's notion of translation and call for a "more ecological approach" (Star and Griesemer 1989, 388). What they mean by this is that they want to move beyond studying translation from the perspective of one single actor, the scientist, to study the simultaneous translations of multiple actors:

> Yet, a central feature of this situation is that entrepreneurs from more than one social world are trying to conduct such translations simultaneously. It is not just a case of interessement from non-scientist to scientist. Unless they use coercion, each translator must maintain the integrity of the interests of the other audiences in order to retain them as allies. Yet this must be done in such a way as to increase the centrality and importance of that entrepreneur's work. The n-way nature of the interessement (or let us say, the challenge intersecting social worlds pose to the coherence of translations) cannot be understood from a single viewpoint.
>
> (Star and Griesemer 1989, 389)

The problem of translation as it is presented here is mainly a problem of collaboration when there is no consensus. This of course is the default case when actors from different scientific disciplines let alone from different sectors of society work together. What Star and Griesemer are interested in is thus the translation/interessement of actors from different social worlds in which objects and methods mean different things. This translation work then requires "substantial labour on everyone's part." (Star and Griesemer 1989, 388).

This ecological view of translation then means that it is not only the practice of *interessement* of one set of actors – that is, their attempts at framing the issue, managing different concerns, and creating an obligatory passage point – that is of interest. Instead, multiple translations are of interest and the whole endeavour of establishing Berkeley's Museum of Vertebrate Zoology becomes the unit of analysis and the question becomes how these different perspectives can be managed and coordinated, that is, successfully translated.

The answer that Star and Griesemer give in their study is that methods of standardisation and the creation of so-called "boundary objects" were key in this translation. Standardised methods with regard to the collection of specimens and fieldnotes proved to be of key importance as they allowed for consistent information and also for collaborating with "researchers at a distance." It can thus also be seen as a mode of governance.[1]

Another noteworthy modification or extension to the notion of translation was provided by Andrew Barry (2013). He explicitly directs attention to the dimension of power and describes translation as a form of *empire-building* in which actors get enrolled into increasingly *durable* relations. As he points out, translation always relates to "movement in space" as much as the "transformation of space (Barry 2013, 414).

From this brief discussion we can already distinguish three basic dimensions of translation that will become useful for the story we want to tell about the RRI work in the three different regions of the TRANSFORM project:

- Translation in its most common understanding means converting a text into another language. The important point then is that this can never be a one-to-one replication, there are always shifts in the meaning between the two texts. Translation thus always involves elements of replication as well as differentiation; *traduction* and *trahison* in Law's terms, alluding to the well-known paronomasy *traduttore traditore*.
- Translation is always political: It refers to the process through which an actor is given, or assumes, authority to speak on behalf of other actors. This is about the enrolment of actors and about the establishment of some actors' problem framings and concerns as the ones shared by a particular group or network. This group or network is built together with establishing these problem framings and concerns.
- Translation also has a geographical dimension, which is inextricably linked to the other two dimensions. Translation implies movement in and transformation of space as the aim is to make particular knowledge claims or representations of reality stable and durable across both time and space. In that sense translation is a mode of "acting at a distance" (Latour 1987) or governing researchers at a distance (Star and Griesemer 1989).

This strand of debate about processes of translation very much focuses on processes of knowledge production. In the next section, we want to expand this discussion by looking at research that focuses on technological artefacts and their travels in what is usually referred to as *technology transfer*.

Translation in studies of technology and organisations

Up until here we have discussed translation in its most common use in linguistics and as an analytic device to study material-semiotic processes of network building. There is, however, another important connotation that the term carries which has to do with a critique of narrow conceptualisations of technology transfer. This strand of debate will become important later when participation and engagement practices are discussed as *technologies* that are supposed to travel across different sites and settings.

Before we dive deeper into this aspect of translation, we need to take a brief detour and ask what we mean when we talk about technologies. While common understandings of what a technology is usually focus on the artefact or device itself, there is a school of thought highlighting the need to think about technology in broader terms.

Allenby and Sarewitz (2013) in that regard suggest thinking about different levels of technology. Using the example of an aircraft, they describe level 1 as the technology or artefact (the aircraft in this example). Level II would be the infrastructure needed for operating this artefact (think, e.g., about flight schedules, and airport personnel). Finally, level III describes the relation of the previous levels to developments on a larger scale (the spread of diseases through increased long-distance travel). This way of thinking about technology also directs attention to the processes in which each novel technology depends on fitting or re-shaping the environments in which they operate, ranging from technical and legal infrastructures to social and moral orderings (Winner 1986; Hecht 1998; Stilgoe 2018).

When thinking about the transfer of technologies, the initial focus of analysis is on the (often implicit) assumptions and premises that go into the design of technologies and on the ways in which these same technologies are then used in sometimes different ways by different actors (de-scription) (Akrich 1992). Engineers inscribe certain ideas about the behaviour of users-to-be into the design of technical objects whereas these users later de-scribe the technology (see also Felt and Fochler 2010). As we will show, what is true for technical objects can also be said of policy concepts like RRI and the various approaches, tools, and techniques attached to it. There are certain principles that are inscribed into this concept by policymakers and (selected) academics, which are then de-scribed in their territorial use. The precise whys and hows of these de-scriptions – their translations – are the subject of this very monograph.

As the notion of technology itself goes beyond the technical artefact to include the networks with which it co-emerges, also technology transfer then becomes broader: It is the translation of such networks. The semiotic idea of *scripts* is in that way brought together with programs of action. In addition, this way of thinking about technology transfer as translation shifts attention from the characteristics of a stable technical artefact to the processes of building and transforming networks and meanings: This then leads to questions about objects and subjects in the making, about the relation between technological change and political systems, and about the relation between technologies and different subject positions:

It is only when the script set out by the designer is acted out – whether in conformity with the intentions of the designer or not – that an integrated network of technical objects and (human and nonhuman) actors is stabilized. And it is only at this point that this network can be characterized by the circulation of a finite number of elements-objects, physical components, or monetary tokens. (...) This is why it makes sense to say that technical objects have political strength. They may change social relations, but they also stabilize, naturalize, depoliticize, and translate these into other media.

(Akrich 1992, 222)

Akrich points to the political nature of technologies and to the question of power relations, questions which are also quite prominent in Barry's concept of *translation zones* (Barry 2013) and which highlight the multiple resistances, barriers, and failures of translation.

A crucial point here also for our work on regional translations of RRI is that strict dichotomies like successful and unsuccessful, or faithful and not faithful uses or descriptions of technologies do not apply here. If we conceive of technologies as more than the object or artefact, then the focus of attention becomes shifts in the networks, practices of tinkering that reshape the original intent behind a certain technology. There might be some ways, in which a technology can be described as successful while it is unsuccessful in other respects. In that sense technologies have been described as "fluid" (de Laet and Mol 2000). They take on many identities, aim to fulfil different purposes simultaneously and in doing so contribute to different networks.

Technologies, understood as shaped by and continuously shaping the contexts or networks of which they become part, are not simply transferred from one setting into another with varying degrees of success. They adapt as much as their context needs to adapt; both are inextricably entwined (Lu and Qiu 2023).

This broadening of ideas about what it means to *transfer* a technology – and in that context also about what constitutes a technology beyond the technical object – has important implications also for thinking about policy terms such as RRI. To further elaborate on this, we will now turn to recent work on the *travel* of techniques for citizen engagement and participation.

Translation and Neo-institutionalism

Recently, the notion of translation has been adopted in the field of organisation studies to describe organisational change through the question of how ideas materialise, travel, and contribute to change processes. In their text "Travels of ideas" Czarniawska and Joerges refer back to Latour to explain why the notion is particularly useful for the study of organisations:

> it means 'displacement, drift, invention, mediation, creation of a new link that did not exist before and modifies in part the two agents' (Latour, 1993, p. 6), that is, those who translated and that which is translated. This explains why the concept is so attractive to us: it comprises what exists and what is created; the relationship between humans and ideas, ideas and objects, and humans and objects – all needed in order to understand what in shorthand we call 'organizational change.'
>
> (Czarniawska and Joerges 2012, 24)

In doing so they contrast translation what they call "the diffusion model" of how ideas spread. While the metaphor of diffusion implies that there is some kind of inertia to a movement and thus some kind of automatic procedure, translation stresses the active work of heterogeneous actors involved in the process. In addition, translation for them is a way to equally address the temporal and spatial aspects of the travel of ideas. Translation used like this clearly goes beyond the linguistic interpretation. It points to similarity and difference but also to the creation of something new: New links between actors that modify these same actors in the process.

The question then becomes why certain translations stick while others don't. Here Czarniawska and Joerges point to the importance of *fashions* in processes of institutionalisation. Fashions for them – similar to what others discuss as *buzzwords* in the policy realm (Bensaude Vincent 2014) – happen at the fringes and margins of the already established and thus have the potential to bring something new. This is of course also interesting when it comes to translating RRI into regional R&I ecosystems. How are certain novel ideas and practices turned into objects (such as linguistic artefacts) and made durable? What are the rationales and justifications that are used here?

This way of thinking about translation in the context of the "travel of ideas" has been applied to the study of engagement activities and the travel of somewhat standardised methods or tools for engagement (Soneryd 2015; Soneryd and Amelung 2016; Laurent 2017; Konopásek, Soneryd, and Svačina 2018).

For our analysis of the work in the TRANSFORM project and its regional clusters, Linda Soneryd's use of the notion for her analysis of engagement activities provides a good entry point. Her work brings together the material-semiotic notion of translation with organisational sociology and focuses on a critical inquiry into "technologies of participation" (Soneryd 2015). Her main interest is in how such technologies get transformed through their practical application in different settings. She explores such processes of translation through examples like scenario workshops, the so-called future workshop, and the citizens' jury approach. What this means is that she is interested in how certain general ideas and normative commitments shift when they are applied in different contexts. This involves ideas about the broader rationales of *empowering citizens* or an aim to *improve decision-making* as well as particular notions of the participants involved. In that way she comparatively traces translations from the original concept to its application in different organisational settings: Is the focus on consensus or conflict, is it about giving voice and representation or about contesting power structures? Are citizens involved to make their needs heard, is it about involving an oppressed citizenry, or are participants understood as epistemic actors and thus as somewhat akin to a scientific citizenry? (Irwin 1995). The important point here is that this is not about whether these translations are *truthful* to the original or not. It is about how shifts and remixes of ideas and approaches – the scripts if you will – coincide with establishing different material links and building new networks.

In that sense, this analytical lens also borrows from research on technology transfer that we introduced above insofar as engagement tools, techniques, and methods are not framed as stable entities but instead are understood as continuously re-shaped:

When public participation instruments are situated in specific local contexts, however, their ideas, values, formal rules, and tools become remixed, giving rise to new meanings.

(Soneryd and Amelung 2016, 171)

Translation as a concept points exactly to such *remixes* and allows for tracing both the symbolic and material elements of such transformations:

Through the concept of translation, it is possible to emphasize not only the epistemic construction of political order, but also its material dimensions.

(Soneryd and Amelung 2016, 172)

This way of thinking about translation focuses on the shifts in the meaning of concepts like participation, citizen, and expert, while at the same time staying attentive to the making and re-making of links between different entities and agents thus directing attention to the organisational settings in which they are applied – or with which they are co-produced.

This way of thinking about translation also takes up the question of what makes certain translations more durable than others? How to create stickiness? Soneryd argues that it is "organizational carriers" and different "normative and symbolic systems" (Soneryd and Amelung 2016, 168) that help stabilise certain translations.

Organisational carriers are thereby understood mainly in the classical sense of organisations and networks. An organisation like the Fondazione Giannino Bassetti, together with the Lombardy Region, adopting the term RRI in their discourse and practice already provides a powerful organisational carrier in this particular region.

When she talks about normative and symbolic systems, this is about taken-for-granted beliefs and unquestioned truths. These can be "ideological frames" or certain institutional "myths."

The durability of different engagement instruments depends on different types of carriers and how they manage to "package" these ideas (Soneryd and Amelung 2016, 168). This understanding resonates with a co-productionist lens in the study of participation and engagement. Similar to Soneryd's argument, such an approach situates collective participatory practices within "wider spaces of participation" and "systemic constitutional stabilities" (Chilvers, Pallett, and Hargreaves 2018). What Jason Chilvers and his colleagues refer to as *wider spaces of participation* consist of particular institutional settings, certain zones of standardisation, and issue spaces. *Constitutional stabilities* include things like legal frameworks, infrastructures, imaginaries, established social practices, and collective forms of public reason. This also points to moments of standardisation like the development of *tools* or *guidelines* that are imagined to be able to *travel* unchanged across different sites and scales, or in the attempts to establish certain groups of actors or communities of practice with the aim to stabilise some coherent forms of engagement and participation. Pointing to such mechanisms, Brice Laurent points out that "what matters is the ability to make the expertise about public participation an expertise about technologies of democracy, separated from the issues on which they are expected to be applied, and which can circulate freely from one issue to the next." (Laurent 2016, 219)

Before we move to the next section, we want to make one more point that will be crucial for our work in TRANSFORM, which relates to a call for research into the performativity of engagement discourses themselves. This is of particular interest as RRI gets increasingly practised as different forms of citizen engagement and participation. In that regard, Soneryd and Amelung argue that

> we need to treat this growing interest in public engagement instruments as a research object in its own right and potentially as a new organized space that changes the conditions for governance.
>
> (Soneryd and Amelung 2016, 157)

This points to questions about the work that RRI discursive practices and implementations are doing in different settings. Who is doing it, and with what consequences? What exactly does it mean when RRI becomes part of Lombardian innovation strategies? What are the consequences of the establishment of a Competence Centre on Participatory and Deliberative Democracy at the European Commission's Joint Research Centre (JRC), and how do the relations between researchers, citizens, and "think tank" members shape the meanings of engagement and participation? These are the kinds of questions that such a framing directs attention to.

Translating ecologies of participation

In the previous section we argued that it might be fruitful to understand the "implementation" of RRI translations into a particular institutional-organisational setting. Such an implementation comes with shifts and remixes of the meaning of RRI. One of these shifts we have observed in recent years is that RRI is increasingly understood – one might even say narrowed down – in terms of engagement and participation in the governance of technoscience and innovation. Thus, before we move back to RRI, it is useful to clarify how we think about engagement practices as part of a broader RRI framing.

To do so, we will take inspiration from recent work on participation and engagement which is situated within a broader literature about "co-production" (Jasanoff 2004). The idiom of co-production states that "the ways in which we know and represent the world (both nature and society) are inseparable from the ways in which we choose to live in it" (Jasanoff 2004, 2). This quote states in a nutshell a more complex set of ideas that basically argue that science, politics, and society are inseparably entwined and even mutually shaping each other. It is also a methodological approach as it asks the researcher studying technoscientific phenomena to become attentive to how scientific, political, social, and moral orderings (which are also always material orderings) are co-emergent.

This approach to studying science and technology has also proven to be useful when thinking about engagement and participation activities. More recently, especially Jason Chilvers and Matthew Kearnes have adopted this conceptual framework and conceptualise engagement exercises as

> contingent and heterogeneous collectives of human and non-human actors, devices, settings, theories, social science methods, public participants, procedures and other artefacts.
>
> (Chilvers and Kearnes 2015, 15)

What is important to note here is that the elements mentioned are regarded as interwoven and mutually constitutive. This means that the public(s) targeted in a certain participation initiative cannot be treated as being separate from the issues that are debated, the cultural and political setting within which they emerge, or the particular engagement *tools* with which citizens are selected and brought into being as a public. That's what is meant when they talk about engagement collectives as being heterogeneous socio-material entities.

Chilvers and Kearnes refer to this understanding as a relational perspective which they distinguish from a more mainstream *residual realist* understanding of engagement. Such an understanding frames engagement exercises as discrete events in which individuals come together in groups to share their pre-fixed opinions on stable issues in unambiguous settings and according to best-practice methodologies. In contrast, engagement from a relational perspective highlights engagement and participation as a set of practices that are entangled with certain techno-political orderings and material institutional-organisational configurations, scientific knowledge claims, objects, issues at stake, subject positions, and (collective) identities as well as particular normative commitments and taken for granted beliefs, that is, institutions such as democracy or responsibility in the governance of technoscience:

> Rather than pre-existing a priori, the subjects (including participating publics), objects (issues or material devices) and models (political ontologies or formats) of participation are actively co-produced through the performance of collective participatory practices [...] which are shaped by (and in turn shape) extant orders on these dimensions.
>
> (Chilvers, Pallett, and Hargreaves 2018, 201)

Such a relational understanding lends itself particularly well to the analysis of different territorial clusters as it is precisely about moving beyond the assessment or evaluation of singular engagement events according to certain quality criteria or indicators. Much to the contrary, it is about going deeper, to carve out the connections between different engagement collectives and their institutional-organisational contexts. To capture such multiple connections between different collectives and practices, Chilvers and Kearnes use the metaphor of "ecology":

> An ecological conception of participation suggests that is not possible to properly understand any one collective of participation without understanding its relational interdependence with other collective participatory practices, technologies of participation, spaces of negotiation and the cultural-political settings in which they become established.
>
> (Chilvers and Kearnes 2015, 52)

Focus on what Chilvers and Kearnes call *ecologies* is deliberately comparative in nature and directs attention to the relations between different engagement collectives and to how they become part of a broader political or democratic culture (which can also be read as an invitation to a more historically inclined analysis).

In other words what counts as political issue and democratic mode of governance is inextricably entwined with ideas about which public need to be involved by whom, in which capacity, and with which agency. Experiments of engagement are neither independent of the institutional and political cultures in which they are carried out, nor are they a mere function of those. Instead, a co-productionist understanding of engagement focused on exploring ecologies of participation aims "to document the specific sites and institutional configurations in which participatory practices cohere and are rendered authoritative" (Chilvers and Kearnes 2015, 53).

As we just alluded to, framing engagement and participation in this way also means re-thinking approaches towards the assessment and evaluation of engagement activities or RRI projects in our case. The main task can no longer be "merely" monitoring and assessing what is done according to the evaluation framework of choice. It is about "staying with the trouble" (Haraway 2016) to carve out the multiplicity and distinctiveness of different approaches and practices while exploring how they are shaped by and shaping broader ecologies of participation and co-produced with particular institutional-organisational, scientific, political, moral, and social orderings. The epistemic stance here is one of learning through difference.

These reflections lead us directly into the final section of this chapter, which we will use to elaborate how the ideas and concepts presented so far will shape the stories we want to tell about the RRI work of the different TRANSFORM clusters.

Summing up: Studying RRI through the concept of translation

I say RRI because it's basically a big, for me/well I know this is recorded but for me RRI is big, how can I say. Like, it's like a big bag where I put a lot of stuff (laughter) in it. But it's all about me being more open, more transparent, involving more the society and I know that there is also the question about ethics and yeah, openness of data and all these things.

(Int_10)

This quote is a nice example of how our colleagues working in the different regional clusters of the TRANSFORM project tend to describe RRI. They point to the multi-faceted and ambiguous nature of the concept. RRI in this instance is compared to *a big bag*. An empty vessel where you can put in *stuff* in that you already are in possession of. There is room for quite a lot of heterogenous things in such a bag. However, our colleague also talks about some things that are constant. She, for example, is guided by a set of principles like openness, transparency, and a stance towards engagement of a broad range of actors. Stretching the metaphor just a little further (hopefully not beyond its breaking point), one could argue that the shape of the bag – its materiality if you will – does matter. While you can put a lot of stuff in there, it is not the case that anything fits. In this sense, this quote points to certain limitations or boundaries. It also points to contingency; it could be otherwise.

Why then does the bag look the way it does (in a particular region) and why do certain things fit better than others?

These are the broad questions that the notion of translation allows us to explore and hopefully find interesting answers to, or which at least provide us with a suitable starting point to tell good stories about territorial RRI projects and practices. It does so by sensitising us to several interrelated themes and issues:

Firstly, translation directs our attention to shifts in meaning. This is already present in the linguistic sense of the term – there is similarity but there is always difference. Translation necessarily involves *betrayal*. These shifts happen when the concept of RRI travels. And travel it does. In a project like TRANSFORM, the idea of RRI (governance) travels from the trans- or supranational European policy world into regional settings. Additionally, it travels between different sectors: From waste management to health, from energy transitions and digitalisation to issues of food waste.

The way translation is used in the debates we described in this chapter focuses on both the shifts and *re-mixes* in meaning, and on the making and re-making of links between different actors. As such the concepts are well-suited to explore the political and institutional-organisational settings in which they are applied. Translation – and this is one of the main advantages of this term – doesn't assume an essence of RRI. Instead, RRI becomes a relational concept in the sense that it undergoes shifts in meaning in relation to other things in the field of activity where it is placed, leading to a displacement and a modification of the concept as well as the new surroundings.

What follows is that as the concept travels and takes on new meanings, it also creates new attachments and gets entangled in different institutional-organisational settings. Thus it is always political in the sense that it is about enrolling actors into a project or into certain ways of doing things. Looking at translations means exploring the ways in which organisational-institutional settings guide how RRI is shaped. It also points to questions about who becomes influential in shaping RRI, whose needs are heard and addressed. Translation thus provides a fruitful way to understand how exactly ideas or policy concepts like RRI – translate and materialise in ever-new forms.

The notion of translation importantly also points us to geographical aspects of how policy concepts spread and move around. The development of standards or even "tools" for RRI that are imagined to be used by various actors in different places and R&I ecosystems implies an idea – or even aim – of controlling actors not only across scale and sectors but also *at a distance*. The idea that people in vastly different settings should do something that is in some way similar and comparable implicates an ambitious drive to govern, which is of course not surprising given that these projects are funded within European funding programs. What is notable, however, is that this resonates with a certain deeply ingrained view of governance itself that has been described already in studies about the emergence of quantification schemes (Porter 1995) and current developments with regard to measuring impact or success[2] which have been described as "measurementality" (Turnhout, Neves, and de Lijster 2014).

In this regard, one important thing that translation does for us is that it allows us to move beyond simplistic dichotomies such as successful/unsuccessful projects or faithful/unfaithful implementations. Unfortunately, these are still common ways to tell stories about RRI projects and their implementation and impact. What we opt for is to "stay with the trouble" (Haraway 2016) and to try to offer detailed descriptions of the work that is done in the different regions and their respective R&I ecosystems while still offering a broader comparative picture of how translations could be otherwise.

This is also where recent developments with regard to a relational view to different forms of participation and deliberative democracy will become important – also as RRI discourses themselves tend to increasingly draw on and get involved in these debates. Talking about RRI in terms of ecologies of participation and engagement collectives stresses the different ecologies in the different regions. Furthermore, looking at the different collectives and ecologies comparatively will enable us to carve out how certain translations become durable within "wider spaces of participation" and "constitutional stabilities." What are the "mediators" and "carriers" that make certain forms of responsibility stick?

Focusing on translation, the stories we want to tell about the work in TRANS-FORM and its regional clusters means looking at how techniques or methodologies of engagement change when they are applied in a particular regional or territorial setting. In TRANSFORM we can assume that this happens on two levels:

- The translation of RRI in Lombardy, Catalonia, and Brussels through the Smart Specialisation Platform (S3 Platform). Which shape does RRI take in the three regions? What is the meaning of responsibility and in which relations is it embedded?
- The translation of certain engagement techniques in these regions: How is *citizen science* translated in Catalonia? What is *design thinking* in Brussels? And how are *participatory agenda setting* and *citizen assemblies* translated in Lombardia?

Hence, we see a double movement of translation: RRI is translated in a specific way in the clusters (through different methods/approaches) which in turn means that also these methods are translated as RRI in specific ways.

One could even make an argument for thinking about RRI itself in terms of *translation*. RRI, when it became influential in European policymaking, allowed actors to bring together and redefine predecessor-initiatives like technology assessment (TA), research on ethical, legal, and soci(et)al implications/aspects of emerging scientific fields and technologies (ELSI/ELSA), anticipatory governance with academic fields like STS and ethics while also managing to weave academic institutions and Directorates-General on an European Commission level into this translation. For around a decade, various actors have been given, or claimed, authority to speak on behalf of other actors about RRI and created a variety of obligatory passage points, with RRI experts as spokespersons. Before we get into the empirical stories about the work of the TRANSFORM clusters, it will therefore be necessary to provide a brief historical perspective on the emergence (and demise?) of RRI. In doing so, we claim authority to speak ourselves on behalf of others about what RRI means, by way of new translations.

Notes

1   Governing actors at a distance is of course also a key concern in European funding programmes and in the evaluation and monitoring of the projects that are funded. Hence it comes as no surprise that standardisation of methods is also a recurring theme in various RRI projects and in the evaluation of their respective success/ impact.
2   One of the instances in which this has been discussed is the discussion paper "Understanding impact, impact pathways and benefits of RRI within SuperMoRRI WP5 and beyond" that was produced by members of the SuperMoRRI consortium. https://super-morri.eu/download/153/findings-and-deliverables/5424/t5-3-and-t5-4-impacts-pathways-and-benefits-of-rri-discussion-paper.pdf. Accessed February 28, 2023.

References

Akrich, Madeleine. 1992. "The De-Scription of Technical Objects." In *Shaping Technology/ Building Society*, edited by Wiebe E Bijker and John Law, 205–24. Cambridge: MIT Press.

Allenby, Braden R., and Daniel Sarewitz. 2013. *The Techno-Human Condition*. The MIT Press. https://mitpress.mit.edu/9780262525251/the-techno-human-condition/.

Åm, Heidrun. 2019. "Limits of Decentered Governance in Science-Society Policies." *Journal of Responsible Innovation* 6 (2): 163–78. https://doi.org/10.1080/23299460. 2019.1605483.

Barry, Andrew. 2013. "The Translation Zone: Between Actor-Network Theory and International Relations." *Millennium: Journal of International Studies* 41 (3): 413–29. https://doi.org/10.1177/0305829813481007.

Bensaude Vincent, Bernadette. 2014. "The Politics of Buzzwords at the Interface of Technoscience, Market and Society: The Case of 'Public Engagement in Science.'" *Public Understanding of Science* 23 (3): 238–53. https://doi.org/10.1177/0963662513515371.

Callon, Michel. 1986. "Some Elements of a Sociology of Translation: Domestication of the Scallops and the Fishermen of St. Brieuc Bay." In *Power, Action and Belief. A New Sociology of Knowledge?* edited by John Law, 196–233. London/Boston/Henley: Routledge/Kegan Paul.

Callon, Michel, and Bruno Latour. 1981. "Unscrewing the Big Leviathan: How Actors Macro-Structure Reality and How Sociologists Help Them to Do So." In *Advances in Social Theory and Methodology: Toward an Integration of Micro- and Macro-Sociologies*, edited by Karin Knorr-Cetina and Aaron V Cicourel, 277–303. Boston: Routledge and Kegan Paul. https://doi.org/10.4324/9781315763880.

Chilvers, Jason, and Matthew Kearnes. 2015. *Remaking Participation: Science, Environment and Emergent Publics*. Abingdon and New York: Routledge.

Chilvers, Jason, Helen Pallett, and Tom Hargreaves. 2018. "Ecologies of Participation in Socio-Technical Change: The Case of Energy System Transitions." *Energy Research and Social Science* 42 (August): 199–210. https://doi.org/10.1016/j.erss.2018.03.020.

Czarniawska, Barbara, and Bernward Joerges. 2012. "Travels of Ideas." In *Translating Organizational Change*. De Gruyter. https://doi.org/10.1515/9783110879735.13.

de Laet, Marianne, and Annemarie Mol. 2000. "The Zimbabwe Bush Pump." *Social Studies of Science* 30 (2): 225–63. https://doi.org/10.1177/030631200030002002.

European Commission, Directorate-General for Research and Innovation. 2015. "Indicators for Promoting and Monitoring Responsible Research and Innovation: Report from the Expert Group on Policy Indicators for Responsible Research and Innovation." Publications Office. https://data.europa.eu/doi/10.2777/9742.

Felt, Ulrike, and Maximilian Fochler. 2010. "Machineries for Making Publics: Inscribing and De-Scribing Publics in Public Engagement." *Minerva* 48 (3): 219–38. https://doi.org/10.1007/s11024-010-9155-x.

Fitjar, Rune Dahl, Paul Benneworth, and Bjørn Terje Asheim. 2019. "Towards Regional Responsible Research and Innovation? Integrating RRI and RIS3 in European Innovation Policy." *Science and Public Policy* 46 (5): 772–83. https://doi.org/10.1093/scipol/scz029.

Haraway, Donna. 2016. *Staying with the Trouble. Making Kin in the Chthulucene*. Durham and London: Duke University Press.

Hecht, Gabrielle. 1998. *The Radiance of France: Nuclear Power and National Identity After World War II*. Cambridge (MA) and London: MIT Press.

Irwin, Alan. 1995. *Citizen Science: A Study of People, Expertise and Sustainable Development*. London and New York: Routledge.

Jasanoff, Sheila. 2004. *States of Knowledge. The Co-Production of Science and Social Order*. London and New York: Routledge.

Konopásek, Zdeněk, Linda Soneryd, and Karel Svačina. 2018. "Lost in Translation: Czech Dialogues by Swedish Design." *Science and Technology Studies* 31 (3): 5–23. https://doi.org/10.23987/sts.65543.

Latour, Bruno. 1987. *Science in Action. How to Follow Scientists and Engineers Through Society*. Cambridge: Harvard University Press.

_____. 1991. "Technology Is Society Made Durable." In *A Sociology of Monsters: Essays on Power, Technology and Domination*, edited by Law, John, 103–31. London/ New York: Routledge.

_____. 1994. "On Technical Mediation." *Common Knowledge* 3 (2): 29–64. http://ecsocman.hse.ru/text/18036068/.

_____. 1999. *Pandora's Hope. Essays on the Reality of Science Studies*. Cambridge/ London: Harvard University Press.

Laurent, Brice. 2016. "Boundary-Making in the International Organization: Public Engagement Expertise at the OECD." In *Knowing Governance: The Epistemic Construction of Political Order*, edited by Jan-Peter Voss and Richard Freeman, 217–35. Palgrave Macmillan. https://doi.org/10.1057/9781137514509_10.

_____. 2017. *Democratic Experiments: Problematizing Nanotechnology and Democracy in Europe and the United States. Democratic Experiments: Problematizing Nanotechnology and Democracy in Europe and the United States*. The MIT Press. https://doi.org/10.26530/oapen_628777.

Law, John. 2003. *Traduction/Trahison: Notes on ANT*. Lancaster: Centre for Science Studies, Lancaster University.

Lu, Miao, and Jack Linchuan Qiu. 2023. "Transfer or Translation? Rethinking Traveling Technologies from the Global South." *Science Technology and Human Values* 48 (2): 272–94. https://doi.org/10.1177/01622439211072205.

Porter, Theodore M. 1995. *Trust in Numbers. The Pursuit of Objectivity in Science and Public Life*. Princeton: Princeton University Press.

Soneryd, Linda. 2015. "Technologies of Participation and the Making of Technologized Futures." In *Remaking Participation: Science, Environment and Emergent Publics*, edited by Jason Chilvers and Matthew Kearnes, 144–61. London and New York: Routledge.

Soneryd, Linda, and Nina Amelung. 2016. "Translating Participation: Scenario Workshops and Citizens' Juries across Situations and Contexts." In *Knowing Governance. The Epistemic Construction of Political Order*, edited by Jan-Peter Voß and Richard Freeman, 155–74. Palgrave Macmillan UK. https://doi.org/10.1057/9781137514509_7.

Star, Susan Leigh, and James R Griesemer. 1989. "Institutional Ecology, 'Translations' and Boundary Objects: Amateurs and Professionals in Berkeley's Museum of Vertebrate Zoology, 1907-39." *Social Studies of Science* 19: 387–420.

Stilgoe, Jack. 2018. "Machine Learning, Social Learning and the Governance of Self-Driving Cars." *Social Studies of Science* 48 (1): 25–56. https://doi.org/10.1177/0306312717741687.

Turnhout, Esther, Katja Neves, and Elisa de Lijster. 2014. "'Measurementality' in Biodiversity Governance: Knowledge, Transparency, and the Intergovernmental Science-Policy Platform on Biodiversity and Ecosystem Services (IPBES)." *Environment and Planning A* 46 (3): 581–97.

Winner, Langdon. 1986. *The Whale and the Reactor: A Search for Limits in an Age of High Technology*. Chicago: University of Chicago Press.

3 Responsibility, RRI, and Science-with-and-for-Society (SwafS)

RRI as a translation

The first decade of the 21st century was an exciting time for those of us who were interested in the governance of technoscience in Europe, that is, the idea of governance put forth in Chapter 1, oriented around ethics, public participation, and new forms of technology assessment (TA). The time was exciting because the idea found its way into the research funding organisations. What happened there, we already explicated in Chapter 2 as acts and forms of *translation.* From the actor's perspective, however, it was often thought of as *implementation,* with all the difficulties, doubts, and qualms that follow. These qualms could be nicely explained with recourse to Pope Francis' encyclical letter *Laudato Si':*

> The basic problem goes even deeper: it is the way that humanity has taken up technology and its development according to an undifferentiated and one-dimensional paradigm. This paradigm exalts the concept of a subject who, using logical and rational procedures, progressively approaches and gains control over an external object.
>
> (Francis 2015, Article 106)

And yet, the excitement of the 2000s was exactly the possibility that the technocratic paradigm itself appeared to be ready for critique and reform. Inspired by the ethical, legal, and social implications (ELSI) component of the Human Genome Project, European countries incorporated ethical, legal, and soci(et)al aspects (ELSA) into funding schemes for research in biotechnology already in the 1990s. Such activities were also funded by the European Union (EU) in FP5, the fifth framework programme for research and technological development (Griessler et al. 2023). Gradually, the ELSI/ELSA community developed ideas about how to move further, to "ELSA 2.0," "integrated ELSA," or simply post-ELSI (Balmer and Bulpin 2013; Balmer et al. 2016), trying to get from a mode where social scientists and humanities scholars operated

DOI: 10.4324/9781003371229-3

as a sort of appendix or apologist for biotechnology and into a more interactive mode where their inputs were integrated into the technoscientific research itself. Ethics, especially in the biotech field, was becoming consolidated, the Oviedo Convention entering into force a month before the turn of the century (or rather, before the turn of the century was celebrated). The remit of research ethics grew, at least in some countries. One example was Norway, where the National Research Ethics Committee for Science and Technology decided that the societal implications of research in terms of their impact on sustainability, democracy, and equity indeed were ethical issues. Parliamentary TA was still going strong, such as with the Danish Board of Technology and what was called Flemish TA (Oudheusden et al. 2015). The many types and strands of science and technology studies (STS) were growing and consolidating, and some strands, such as the scholarship and practices labelled post-normal science (Funtowicz and Ravetz 1993), were embraced and adopted well beyond STS academe, see for instance Castro e Silva and Teixeira (2011). On top of this, new fields of so-called converging or emerging technologies were developing, above all nanotechnology and synthetic biology, providing scholars and practitioners with new funding opportunities as well as ample ground for study and action.

EU institutions played an important role in this development. EU's sixth framework programme (FP6; active years 2002–2006) included a separate programme called Science and Society with a budget of 88 million euros for projects and activities to support the study and practical implementation of ethics, TA, studies of risk and precaution, science-society dialogue, and related fields. Indeed, already the FP6 Science and Society programme expressed the need for *responsible research*, a better dialogue between Science and Society, and better alignment of research agendas with the concerns of civil society. The seventh framework programme, FP7 (2007–2013), developed this further into "Science in Society," trebling the investment to 330 million euros to be spent over the 7 years FP7 period. In 2000, the European Commission (EC) published its Communication on the Precautionary Principle, an (arguably slow) follow-up of the Rio Declaration, followed in 2001 by the much-cited report by the European Environment Agency called "Late Lessons from Early Warnings." In this report, the agency presented numerous case studies of unforeseen environmental and health hazards created by technological developments and argued for the need for new governance approaches based on precaution. The same year, the EC published one of its more remarkable statements, the White Paper on European Governance (see Chapter 1) which highlighted the need for public participation also in the context of science and thereby provided a justification for the Science and/in Society programmes. As FP7 developed, the question of how to develop what was called a more dynamic governance of science and technology and the aspect of responsibility were becoming ever more prominent in the Science-in-Society programme.

The concept of responsibility itself could be seen both as a translation and a boundary object. The concept is old and belongs to everyday language as well as various institutional contexts, not the least legal contexts. "Responsible research" is a term that probably every actor in the research and innovation (R&I) system could fill with content, on a range from methodological rigour and research integrity to issues of the societal accountability and applicability of results and beyond. At the same time, within the FP7 Science in Society programme and the academic discourse that grew around it (partly because of the funding opportunities that the programme created), the term responsibility came to represent a value that was defined in terms of what those outside the combined boundary of STS, ethics, TA, and philosophy of technology rightly could characterise as esoteric discourse. "Responsible research" came to mean research that exerts responsibility vis-à-vis society by taking into account, responding to, and aligning to needs and concerns expressed by civil society and its citizens. A loose thought collective emerged around this concept of responsibility and its main proponents, in particular René von Schomberg who played a dual role as a key contributor to the philosophical discourse and as a *fonctionnaire* and policymaker in DG RTD, the European Commission's Directorate-General for Research and Technology Development, later Research and Innovation. This duality or ambiguity between the colloquial, common-sense understanding of responsibility and the emerging theory-laden concept of responsible research provided opportunities for initiatives that introduced responsibility into R&I policy. Who could be against responsibility? And at the same time, in possibly the most esoteric of policy pieces on responsibility and the governance of technoscience, the EC recommendation on a code of conduct for responsible nanosciences and nanotechnologies, responsibility is defined in terms of a set of seven principles: Meaning, sustainability, precaution, inclusiveness, excellence, innovation, and accountability. The paragraph of "meaning" was remarkable: "Nanoscience and nanotechnology research activities should be comprehensible to the public. They should respect fundamental rights and be conducted in the interest of the well-being of individuals and society in their design, implementation, dissemination, and use."

In Callon's terms the choice of the term "responsibility" could be seen as a translation by which the diverse and uncoordinated traditions – ethics, TA, public participation, etc. – of trying to govern technoscience got a common umbrella concept. Indeed, it offered *interessement,* a coordination and stabilisation of hitherto scattered thought collectives that now could mobilise around a shared goal. The term responsibility was well known, bore connotations in the right direction, and yet was empty enough to avoid immediate controversy. One observation is that with the exception of the nano code, the issue of sustainability was not frequently invoked in the context of responsible research; for instance, it never got to be one of the "RRI keys." It is tempting to speculate that it would signal a too substantive value that indeed would show tension with the dominant discourse of the primacy and necessity of innovation for economic growth. On the other hand, it could be argued that the EU emptied also that signifier around the same time, by introducing the idea of "sustainable growth." The slogan of the "Europe 2020" policy was even "smart, inclusive and sustainable growth," the EU equivalent of a Kinder Surprise.

Viewed from the perspective of translation, however, the concept of responsibility is perhaps too colloquial and too much present in other contexts to allow for precise problematisation and mobilisation. When everybody already has an idea about what responsibility is and should be, the signal-to-noise ratio is difficult to improve. Responsibility was not a new concept that entered into an empty space (Kjølberg and Strand 2011). However, when the acronym of RRI began to appear at the beginning of the 2010s, it offered novelty and mnemotechnical distinctiveness. Indeed, when working on this chapter, we were surprised to discover that both the FP6 Science and Society and the FP7 Science in Society work programmes explicitly and prominently used the term "responsible research." We ourselves had either forgotten or never noticed, in spite of having written several research proposals and received funding from the Science in Society programme.

RRI, SwafS, and Horizon 2020

After summer comes winter. There are several chronicles and analyses of the emergence of the RRI acronym and its fate within EU framework programmes (Owen, Macnaghten, and Stilgoe 2012; Macq, Tancoigne, and Strasser 2020; Strand and Spaapen 2021; Daimer, Berghäuser, and Lindner 2023; Griessler et al. 2023). Most of them tell a story of rise and decline in the 2010s, beginning with the entry of Robert Jan Smits as the new Director-General of DG RTD in 2010 and his failing interest if not outright hostility towards public participation in science in general and the FP7 Science in Society programme in particular. It appeared that Smits understood what many civil servants and politicians did not fully grasp, namely that the philosophies and policies of Science and/in Society, of public participation, ethics, good governance, etc., were based on a critique of modernity and an ambition of a new social contract of science. We may guess that from Smits' perspective, in a Europe struggling to recover from the 2008–2009 financial crisis, such critique appeared as sand in the machinery of innovation and growth and a luxury that could no longer be afforded (Macq et al. 2020).

As Director-General, Smits immediately proceeded to reorganise DG RTD. One of the results was that the Science and Society policy field was heavily downsized and lost its own separate directorate within DG RTD, a process of decimation that continued steadily when Carlos Moedas was made Commissioner for R&I in 2015. At the time of writing (2023), there is virtually nobody left in DG RTD with a mandate to work on these issues. At a distance this may sound paradoxical, given that EU's eighth framework programme, Horizon 2020 (active years 2014–2020), featured RRI as a cross-cutting principle that was supposed to be mainstreamed into all EU research, as well as a new, quite large, and ambitious unit of the programme called Science-with-and-for-Society (SwafS). The short story is that the entry of RRI and SwafS into Horizon 2020 was the result of successful politicking against Smits from the side, from above (the European Parliament), and from below. The remaining DG RTD staff in the area gathered efforts and creatively launched "RRI" to salvage the increasingly marginalised policy area, translating not only the above-mentioned intellectual traditions into a discourse on responsibility but also most of the actual work areas of the previous Science and Society directorate into what became known as the RRI "keys" of ethics, gender, open access, science education, and public engagement (and sometimes, the sixth key called governance) (Macq et al. 2020). In parallel though definitely not in harmony with these keys, René von Schomberg developed the philosophical justification and definition of RRI as the "transparent, interactive process by which societal actors and innovators become mutually responsive to each other with a view on the (ethical) acceptability, sustainability and societal desirability of the innovation process and its marketable products" (von Schomberg 2011, 9). Somehow, the RRI concept was able to catch traction all the way up to the level of the Commissioner at the time, Máire Geoghegan-Quinn (Owen et al. 2012), perhaps for the reason explained above: It worked as a boundary object in the sense that it appeared meaningful and reasonable even to the political laity who knew nothing about STS and critiques of modernity.

Furthermore, the idea of a sequel to Science in Society was developed. It got the cumbersome name of Science-with-and-for-Society, coined with a narrative of a logical development from discovering the importance of the relationship between Science and Society (FP6-SaS), moving through the realisation that they are entangled (FP7-SiS) and finally affirming that science should be aligned with and in the service of society (the SwafS programme). STS scholars and other academics and practitioners who believed in these ideas and admittedly also were in danger of losing an important funding stream, created lobbying initiatives in several European countries and allied with European parliamentarians. In the end, the European Parliament and the Council instructed the EC to create SwafS, actually not as a programme but an entire unit of Horizon 2020 with an impressive budget of 462 million euros.

To some of us who were involved in these developments, there was already at the time – 2013 and 2014 – a growing suspicion that the SwafS programme might be a Pyrrhic victory. Whether it was so or not, the ambitions of a EC recommendation on RRI, along the lines of the nano code of conduct but applicable to all scientific fields, dismantled and eventually landed on the so-called Rome Declaration of RRI, inspired by the Lund Declaration on Grand Challenges but never living up to the importance of the latter. The ambitions of mainstreaming

RRI into the entire Horizon 2020 by incorporating the principle into other work programmes and calls fizzled out in the absence of support from above in the DG RTD hierarchy. As for SwafS itself, Arie Rip (2016)'s essay about his work in its initial expert advisory group is a telling story about his encounter with an organisation that in no way invested whole-heartedly into the creation of a strong and competent programme. Rather, the policy staff for this new half-a-billion-euro investment was even further reduced in what was difficult to interpret as anything else but an act of internal correction and retribution for us who were bystanders.[1] Yet, the programme, or rather unit, of Horizon 2020 came into existence and operation and funded in all 261 "actions," that is, activities and projects, until the area was practically eliminated in the ninth EU framework programme, the co-called Horizon Europe, as was the concept of RRI itself.

Some of the analyses of what we might call the rise and fall of RRI in the EU focus on the political and institutional explanations of the fall, also with an eye to what could have been done differently. Griessler et al. (2023) emphasise the conceptual and institutional fragility of RRI. Loeber, Bernstein, and Nieminen (2023) analyse and criticise the New Public Management (NPM) character of its implementation. Along similar lines, Daimer et al. (2023) interpret the fall as a failure of deep institutionalisation.

While these analyses are instructive and relevant for learning, our focus is another one. Firstly, as was highlighted in Chapter 1, the introduction of the concept of responsibility and the policy principle of RRI can be seen as an attempt to translate a vision for governance of technoscience based in a critique of modernity, that is, a vision that both practically and philosophically engages with the idea that modern institutions – and science and technology *per excellence* – contribute to harm, risk, inequality, and injustice as well as their more celebrated effects. In that light, it is hardly surprising to experience a pushback in those same institutions. Put simply, when a policy requires something to be done in order that research becomes responsible, those who created that policy can only with difficulty deny that there is an implicit claim of research currently being irresponsible. For some, such a claim is offensive; for others, it is close to unthinkable.

Rather, it was an impressive if not astonishing achievement to push the concept up to the level of Commissioner speeches, funding programmes, and the Rome Declaration in the first place. RRI, as it was conceived, was and is at odds with what Pope Francis later came to refer to as the "one-dimensional paradigm" in his encyclical letter, and that technocratic, capitalist, and consumerist one-dimensional paradigm was the dominant one in EU policymaking as indeed in the entire Organization for Economic Co-operation and Development (OECD) area and beyond. The so-called science wars showed that such critique can become the target of pushback and attack even when relegated to obscure journals and university seminars, especially when economic interests are perceived to be at stake (Hilgartner, 1997). In the EU, the 2008–2009 financial crisis led to a reorientation of EU research policies with an emphasis on innovation for job creation and economic growth, and a concomitant change of DG RTD leadership as described. However, even without this contingency, it is difficult to imagine that RRI would not have clashed, sooner or later, with the vested and ideological interests in a modernist conception of science as a benefactor of society, providing truth to power and prosperity and wellbeing by the linear model of innovation. The stakes in this issue are larger than what can be managed by clever tactics.

The opening statement of Steven Shapin's (1998) book about the Scientific Revolution reads: "There was no such thing as the Scientific Revolution, and this is a book about it." Shapin argued that if time travel had been possible and we could have visited the 16th and 17th centuries to meet Galileo and his like, we would not have heard them speak of a scientific revolution. This revolution is a historical construct to be appreciated for us who live three and four centuries later, at dusk, as the owl of Minerva spreads its wings and flies. Only in hindsight can one understand a development that took several generations and only now can we trace its effects. Similarly, one should not underestimate the size of the challenge of governance of technoscience and the time required for it. RRI in the fullest sense, in von Schomberg's conception of a new governance model for a science and technology that has become a runaway train, is a matter of leaving behind naïve optimism in science as an unproblematic source of progress, one of the few such sources left in a world struggling with its grand narratives. How modern societies will pull it through into reflexive modernisation and subsequently into sustainability and if a more responsible governance of technoscience will play a role in those transitions, may be a question for future longue durée historians. That is, if there will be such historians.

To the extent that they will exist and at all will be interested, one of the things the historians might wonder about is the intensity of the attempts made in broad daylight and in real time to chronicle and assess RRI initiatives, often by individuals who simultaneously act as chroniclers and practitioners/policymakers/ activists. The first historical paper on RRI was published a few months after the concept made its first appearance on the policy scene (Owen et al. 2012). A striking feature of the so-called RRI scholarship is that several of the main protagonists of RRI – notably René von Schomberg, Richard Owen, Jack Stilgoe, and Phil Macnaghten – not only combined RRI scholarship with being RRI actors but appeared to actively use the scholarship, including the ability to act as their own chroniclers, as part of the efforts to promote RRI. Shanley (Shanley, 2022) describes this as a more general phenomenon in the field and sees RRI as an example of a "scientific-intellectual movement." The recent anthology from the NewHoRRIzon project – one of the RRI flagship projects funded by the SwafS programme – largely confirms her claim as it includes several chapters in the genre of contemporary history and analysis (Griessler and Blok 2023).

In part placing ourselves and this book within that genre, we would like to make three observations. Firstly, what is being exerted in this genre could be seen as the type of reflexivity in which STS and philosophy excel. These are not the only fields within the social sciences and the humanities in which self-awareness and self-reflection are highly valued methodological virtues. Still, especially in STS the virtues of reflexivity and self-reference have been cultivated at a level bordering to obsession ever since the so-called chicken debate in the 1990s (Pickering 1992). Secondly, on a much more practical level, a lot of the so-called RRI scholarship is output from what rightly is called "actions" in the EU funding system. In the course of the science and/in/with-and-for society programmes, it is true that a sizeable number of research projects proper were funded but many projects were of other types, namely so-called support actions and coordination actions. We shall return to this point to discuss how that organisational choice shaped the space for possible translation of RRI in terms of its enactment. The choice also had consequences for scholarship, however. For researchers involved in such actions, there is the need to maintain one's own academic capital by scientific publishing. At the same time, the funding is formally speaking granted for other work than academic research. One solution to this imminent career problem for SSH scholars is to write academically about what one has been doing and what happened in the project and thereby "get publications from the project," as it is called in everyday conversations. Hence, publications may acquire a navel-gazing flavour simply because of work conditions.

Our third observation is perhaps more analytical and speculative. Future historians might be struck by the degree of normativity of RRI scholarship in the 2010s and 2020s, as with the ELSI/ELSA scholarship in the decades before. The literature abounds with proposals about what RRI ought to be and how it ought to be done (Fisher 2005; Davies, Kearnes, and Macnaghten 2010; Delgado and Åm 2018) with chronicles about what was done and if it was a success or a failure (Barben et al. 2007; Nordmann 2007; Åm 2019); and with numerous suggestions about how to monitor and assess the qualities and achievements of RRI work (Yaghmaei and van de Poel 2020). Successes and failures as well as recommendations for good or even best practice are made on the basis of quite few experiences, perhaps just one single project, perhaps already at the moment the project closes. Again, this is in part due to the organisation of the funding, and in particular the requirement to "assess impact," as if this can be done in real time in a methodologically meaningful way. Loeber et al. (2023) criticised that RRI was implemented according to a NPM type of logic. That may be true, but perhaps the "scientific-intellectual movement" of RRI itself embodied a sense of urgency and a degree of impatience characteristic of NPM. Perhaps this movement itself implicitly believed that RRI was a matter of fixing science, that the matter was urgent, and that the question of whether efforts had the desirable effect, could be answered in the short term and without too much complexity. As RRI scholars-practitioners ourselves, we have witnessed how scientists who were the target of RRI actions or policies, spontaneously thought of RRI as yet another form of NPM bureaucratisation. It is worthwhile to reflect on the question if not RRI as a particular vision for governance of technoscience itself emerged within and was shaped by what the Bishop of Rome called the one-dimensional paradigm.

We shall not pursue this speculation to greater length in this chapter. What has had bearing on our own work, however, is our doubt about attempts at measuring or assessing results of RRI efforts in the short term. If the way out of the one-dimensional paradigm is a matter of longue durée and not a quick fix, one should think that there is a need to slowly document and understand what is being done rather than taking for granted the actor's perspective. There is a need for academic distance to understand, a need that is always there in social science but quite acute in the RRI field for the three reasons we outlined above. Indeed, viewing these normative reflections on RRI through the lens of translation may help us understand them at least partly as efforts by the RRI research community to (re-)claim the authority to speak on behalf of responsibility, to tame the Leviathan so to speak (Callon and Latour 2006), while responding to criticism and push-backs.

This book does not entirely meet that need. It is still an example of chronicling by RRI scholars-practitioners. The book offers two qualities, though. Firstly, the author team, while definitely a part of the project that we are about to introduce, the Horizon 2020-funded coordination and support action TRANS-FORM, was positioned at some distance from the shop floor action because our work was organised in a separate work package called "Monitoring and Evaluation." Secondly, in our empirical descriptions, we try to provide thick descriptions of the practices as they developed rather than mainly comparing them to the normative blueprints by which they were surrounded. Our aim was to describe and understand what happened when RRI, which we have described as a translation of certain ideas and efforts, itself was translated into the SwafS programme. Which problems, goals, roles, and voices came to the fore? What did RRI become in Horizon 2020? This is the question that we pursue in this and the following chapters, moving step by step from the SwafS work programme to our case study of a call text, a proposal, a grant agreement, and finally a set of activities funded by the Horizon 2020 programme. Before we get there, however, it is useful to take a sidestep and briefly look at some ways in which RRI did not get to be translated.

The RRI that never was (yet)

When trying to understand the trajectory of RRI through EU R&I policy, there is very little need to engage in counterfactual history. The actual sequence of events already offers enough data points to compare facts from foil. The most readily available foils are the expert reports and policy initiatives that failed to be translated into action in the policy cycle for which they were intended. We believe such comparisons can give valuable hints about conditions of possibility as well as necessity in the process whereby a new concept such as RRI travels from policy to operationalisation.

The perspective taken in those comparisons is that of short-term direct translation. Connoisseurs of STS are likely to agree that the more sophisticated expert reports of the period that we are discussing, all had in common that their immediate policy uptake was close to zero. Instead, they live a longer life as inspirational reading for scholars and presumably also policymakers. For instance, according to Google Scholar, the citation frequency for *Taking the European Knowledge Society Seriously* has been more or less constant from its publication in 2007 until now (2023). *Late Lessons from Early Warnings* was published in 2001 and was most cited in the decade 2010–2019. *Converging Technologies: Shaping the Future of European Societies* is still present at least in academic debates. If indeed we are right that RRI is part of a long-term challenge for modern societies, it is definitely possible that the theoretically more profound contributions ultimately also will be the more impactful ones by becoming standard references or even classics. Perhaps then our successors are going to see that there was a revolution.

Getting closer to RRI, we will briefly present two outputs each from two lines of work that did not materialise into action in the short term. The first line is connected to the already mentioned René von Schomberg, who was a policy officer in DG RTD and at the same time a significant contributor to the philosophy of technology in this field. The second pair of outputs is connected to another combined expert-policy officer of the period, namely Lino Paula, a *fonctionnaire* in DG RTD who had his own research career within the ethics and public engagement of technoscience and held a PhD degree in STS.

von Schomberg is well known for his definition of RRI and the philosophical justification he provided for it, based in his reading of the Collingridge dilemma and his narrative of the historical development of governance of science and technology. He did however also launch several concrete suggestions for practical implementation. The nano code, or the EC Recommendation on a code of conduct for responsible nanosciences and nanotechnologies, is a EC document and as such the result of interactions between numerous individuals. The official document itself carries the name of another DG RTD policy officer, Philip Galiay (see also Galiay 2011), whose importance for RRI in SwafS hardly can be overestimated as the policy area fragmented and von Schomberg, Paula, and several others left or were transferred to other areas after 2013. However, von Schomberg's major contribution to the nano code can be confirmed by reading the 2007 EC working document that was authored by him and that largely anticipated the nano code (von Schomberg 2007).

The nano code does not use the acronym RRI but is otherwise as RRI as it gets. It was finalised in 2008, that is, pre-Smits. Its core idea is that of encouraging "soft governance" in the sense of voluntary deliberation and reflection processes among stakeholders, and then providing principles to be considered in the deliberation. The principles were mentioned above: Meaning, sustainability, precaution, inclusiveness, excellence, innovation, and accountability. In oral presentations, the set of principles was presented as a kind of menu, in the sense that stakeholders were thought to choose the ones that they found relevant in a given setting. The full set of principles contained tensions or even contradictions, but such tensions were also in a sense a reflection of reality.

We are unaware of evidence of direct use of the nano code other than in education and in scholarship studying nanoscientists and nanotechnologists (see, e.g., Kjølberg and Strand 2011; Glerup, Davies, and Horst 2017). In fact, what DG Industry had to say about nanotechnology in 2013, is glaringly anti-RRI, with headlines of the type "How nanotechnology will help Europe make everything better" (European Commission 2013a) and passages that called for efforts to make the public accept the new and wonderful technologies. The nano code did not become mainstreamed.

During the following years, von Schomberg continued his conceptual work on how to structure deliberation on science and technology, and our next example is his RRI matrix, see Figure 3.1 (von Schomberg 2013).

The underlying idea of the matrix was that it connects the practical question of how to organise collective responsibility with what von Schomberg called the normative anchor points that provide RRI with its institutional justification within the EU. The anchor points were derived from the Treaty on the EU; a similar strategy was sought with the Rome Declaration on RRI, which fetched its normative anchor points from the EU Charter on fundamental rights. As far as we know, the table caught some interest in the circles of RRI scholarship but not beyond.

Another intellectually ambitious attempt of the time is the report of the Expert Group on the State of Art in Europe on RRI, chaired by Jeroen van den Hoven and supported by Lino Paula as the policy officer (European Commission 2013b). The report explored various options for the institutional embedding of RRI. Its recommendation was to support the integration of RRI not only in EU funding but also in public and private research funding in EU member states, by improving policy coordination but also by developing mechanisms for training and awareness raising. Also in this case, the advice was not much listened to and especially not by the new leadership of DG RTD. The complexity of the advice probably did not help. Indeed, the report may be best described as a thorough and dense document. We include as Figure 3.2 a facsimile of its Annex I, called "Definition of RRI" as an indication of the genre. Its definition of RRI counts 1,780 words.

Along similar lines, the Expert Group on Policy Indicators for RRI, chaired by one of the authors of this book (Roger Strand) and also supported by Paula, set out to solve the problem of how to satisfy the control-and-command and

Product-dimension → / Process-dimension ↑	Technology assessment and foresight	Application of the precautionary principle	Normative/ethical principles to design technology	Innovation governance and stakeholder involvement	Public engagement
Technology assessment and foresight	–	Development of procedures to cope with risks	Which design objectives to choose?	Stakeholder involvement in foresight and TA	How to engage the public?
Application of the precautionary principle	Identification of nature of risks	–	Choice and development of standards	Defining proportionality: how much precaution?	How safe is safe enough?
Normative/ethical principles to design technology	"Privacy" and "safety" by design	Setting of risk/uncertainty thresholds	–	Which principles to choose?	Which technologies for which social desirable goals?
Innovation governance models and stakeholder involvement	Defining scope and methodology for TA/foresight by stakeholders	Defining the precautionary approaches by stakeholders	Translating normative principles in technological design	–	How can innovation be geared toward social desirable objective
Public engagement and public debate	Defining/choice of methodology for public engagement	Setting of acceptable standards	Setting of social desirability of RRI outcome	Stakeholders roles in achieving social desirable outcomes	–

Figure 3.1 The RRI matrix as developed by Rene von Schomberg (2013).

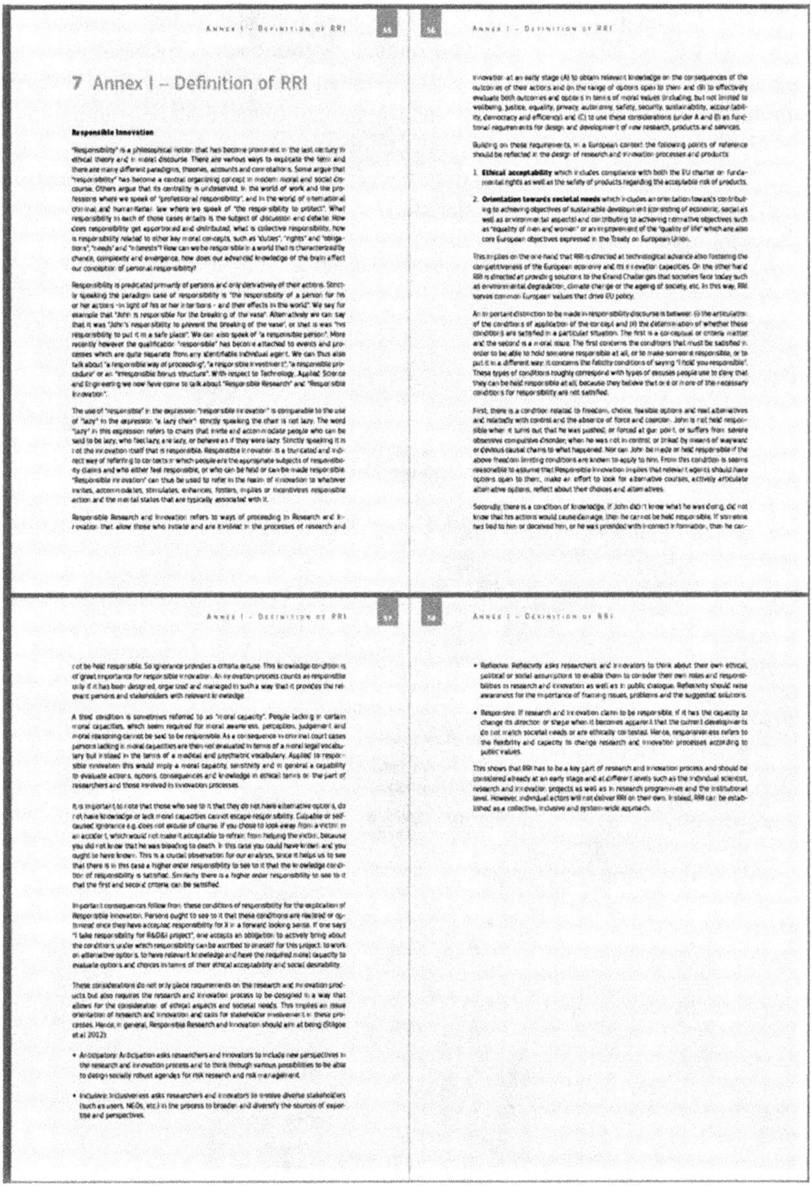

Figure 3.2 RRI definition from Annex I of the report of the Expert Group on the State of Art in Europe on Responsible Research and Innovation entitled "Options for strengthening responsible research and innovation."

NPM-like expectations of "SMART" indicators for each of the so-called RRI keys that by then (2014) had been introduced, without betraying basic values of RRI such as transversal dialogue, soft governance, and deliberation (European Commission 2015). The result was once again complicated and encumbered by tensions – we include as illustration the table of proposed indicators for public engagement (Figure 3.3). By the time the report was approved for publication, the members of the expert group already knew that it would have little or no life in ongoing policy cycles.

Criteria	Performance indicators		Perception indicators	Key actors
	Process indicators	Outcome		
Policies, regulation and frameworks	Formal commitment	PE funding percentage from R&I Public influence on research agendas Share of PE in R&I projects based on consultation, deliberation or collaboration	Public expectations of involvement Researchers' openness to pursue PE Interest of publics	
Event and initiative making Attention creation	Science events and cycles Referenda and Danish-model activities. Organised debates Museums/science centres Informal settings Citizen science initiatives Crowdfunded science and technology development	Media coverage Social media/web 2.0 attention Museum visits and impacts (on visitors, stakeholders, local communities) Civil society organisation activities and impacts	Engagement activities (ladder) Interest in science Issue discrimination Image of an 'atmosphere' of scientific culture	• States • Regions • Cities • Universities • University departments • Research centres • Research projects • Sections of the public • Civil society organisations
Competence building	Training of communicators Training of scientists/engineers Mediators Grass roots	PR staffing Social scientists collaboration In-house/outsourced consultancies The state of science journalism	Knowledge, beliefs Trust, confidence Attitudes (utilitarian expectations, fundamental orientations)	

Figure 3.3 Table on RRI indicators from the report of the Expert Group on Policy Indicators for RRI.

Enter the six (or five) keys

Bearing in mind the foil of initiatives that did not get to implemented in their current policy cycles, we shall now turn to the actual translations of RRI. We shall remain within the context of the eighth EU framework programme, the Horizon 2020. This is not to say that there were no other interesting translations of RRI. On the contrary, the so-called AREA[2] framework for responsible innovation of the UK Engineering and Physical Sciences Research Council was in many ways closer to the academic ambitions, especially of Owen, Stilgoe, and Macnaghten, and it influenced RRI work in other countries. In Chapter 8, we provide a contrast to SwafS by discussing how the AREA framework was adopted by the Research Council of Norway. To some extent, the AREA framework also played a role in several of the projects and initiatives that belonged to the SwafS programme.

Still, the main route for RRI in Horizon 2020 was not that of AREA but of the six keys – public engagement, gender equality, science education, open access, ethics, and finally governance, which in the original leaflet that introduced the keys, was described as "the umbrella for all the others" (European Commission 2012). The leaflet is a prime example of the first translation of RRI from, as it were, von Schomberg's philosophical analysis to a policy concept that could somehow coexist with and inside Horizon 2020. Its title is "Responsible Research and Innovation: Europe's ability to respond to societal challenges," thus placing RRI within the "(grand) challenges" discourse that got increasing traction in EU R&I policy during and after the financial crisis in 2008–2009 (European Commission 2008) and in particular with the Lund Declaration. Indeed, the largest unit of Horizon 2020 was the one called Societal Challenges. In the EC leaflet, RRI was positioned as a facilitator:

> Since 2010 the focus of SiS has been to develop a concept responding to the aspirations and ambitions of European citizens: a framework for Responsible Research and Innovation (RRI). The grand societal challenges that lie before us will have a far better chance of being tackled if all societal actors are fully engaged in the coconstruction of innovative solutions, products and services.
>
> (European Commission 2012, 2)

The text continued by explaining how each of the RRI keys would make a contribution: Public engagement in order to "develop joint solutions to societal problems and opportunities" and – and this is again a direct quote: "to pre-empt possible public value failures of future innovation." Gender equality was introduced not as a matter of discrimination or fundamental rights but as a matter of "unlocking the full potential." The train of innovation and growth can go faster if the engine is fuelled with both men and women, and the existence of SwafS in

Horizon 2020 can be defended because it will be a lubricant for the other units of Horizon 2020. In this way it was possible also to justify the inclusion of the possibly most surprising of the RRI keys, namely that of science education:

> Europe must not only increase its number of researchers, it also needs to enhance the current education process to better equip future researchers and other societal actors with the necessary knowledge and tools to fully participate and take responsibility in the research and innovation process. There is an urgent need to boost the interest of children and youth in maths, science and technology, so they can become the researchers of tomorrow, and contribute to a science-literate society.
>
> (European Commission 2012, 3)

Strand (2019) argued that RRI debates never get to consensus on what RRI is in part due to the unclarity and ambiguities about what it is for. For some, the problem for which RRI is the solution can be stated (arguably in an exaggerated manner) as follows: "How do we regain control over the runaway train of science and technology before it totally destroys our world?" On the other extreme and possibly caricatured end of the axis we find the problem "How do we educate, reassure and calm down the ungrateful public and make them trust us, trust science again?" What one can observe in all official EU documents that explain RRI, from the Horizon 2020 framework itself to the mentioned 2012 leaflet and further in the SwafS work programme, is the art of accommodating and balancing versions of both policy narratives. The critique of modernity is there, albeit in implicit forms, such as when RRI is defined in terms of better alignment of R&I with societal values, needs, and concerns. In that way, Horizon 2020 implicitly calls for society to speak back to science and for science to listen. When this is to be broken down to something more concrete such as the keys, however, signs of technological optimism and the information deficit model always show up. This is how, of all things, science *education* can become a key of RRI. And more than often, the formulations keep enough unclarity and ambiguity in order for RRI to be able to work as a boundary concept, giving some space for the few expert policy staff who were left in DG RTD to be able to operate the SwafS programme while not upsetting the innovation-for-growth leadership of DG RTD too much. One has to admire the political ingenuity of crafting phrases such as "to pre-empt possible public value failures of future innovation." For the outsiders, it has the comfortable flavour of the information deficit model; for the insiders, it sends a signal, albeit a weak one, that there still is some critical edge left. And finally, the keys meant that work areas within the former Directorate for Science and Society could be kept, such as actions directed towards youth. This was the concrete reference of the science education key; it was clearly not about an educational reform. RRI belonged to DG RTD and not to DG Education and

Culture. The keys represented in this sense continuity, in that policy officers could continue to create and promote types of calls and funding streams in the SwafS programme that could be recognised as similar with their predecessors in FP6 and FP7.

The SwafS work programme

There is abundant RRI scholarship on the difficulties of "implementing" RRI by institutionalising RRI practices through an EU project funding scheme (Rip 2016; Ribeiro et al. 2018; Owen, von Schomberg, and Macnaghten 2021). Wicher and Frankus (2023) studied the traces of organisational learning and change in one particular SwafS project funded under a call entitled "Supporting structural change in research organisations to promote RRI." They concluded:

> "Our example shows again that the change of organisations towards a more responsible research cannot be implemented with this key-related thinking that is squeezed into a three-year project."
>
> (Wicher and Frankus 2023)

In what follows, we will continue to try to avoid the normative perspective of (successful or unsuccessful) *implementation,* which indeed is the conceptualisation that is closest to the funders and possibly the project participants themselves, and rather take the more descriptive perspective of *translation.* In that light, we can follow the development of RRI in the successive SwafS work programmes for the years 2014–2015, 2016–2017, and 2018–2020. We shall now leave the genre of chronicling Brussels events of negotiation and politicking and rather stay close to the texts of the work programmes themselves as the very literal translations of RRI in Horizon 2020 (given that the mainstreaming that took place within the Societal Challenges part of the framework programme were rather marginal).

The consecutive SwafS work programmes are first of all not extremely different from FP7 Science in Society work programmes. In all of them, one finds the general policy narrative that balances the critical aspect – the so-called philosophical RRI that calls for better governance of technoscience – with the need to conform to the dominant fostering-innovation-for-growth policy discourse that dominated the period. The duality is expressed through juxtapositions, ambiguities, and occasional cognitive dissonance, as one would expect. Moreover, we find several of the funding streams that were there from previous framework programmes – calls for gender equality actions, calls for initiatives directed towards youth and their interest in science and technology, and calls for project in ethics as well as more conceptual projects on governance of science and technology.

Secondly, one can easily trace the introduction and disappearance of "hot" policy concepts in the work programmes. In the 2014–2015 version, "co-creation" appears, while the 2018–2020 version adds "transdisciplinary multi-stakeholder approaches," which in the meantime had got a rising star in the OECD and then the EU.

As documented by Rip (2016), the SwafS programme felt its way forward and perhaps had to. The introduction to the first work programme is very sparse and cautious in its wording – it is one and a half page long, mainly rehearsing the argument that the RRI keys will foster innovation and outlining how the set of calls – most of them named after keys with a set of technical codes (SEAC, GERI, GARRI) that disappeared again with the next work programme – connected to that goal. In the 2016–2017 work programme, the introduction is somewhat braver and speaks of its 8 (!) keys, where "scientific careers" now is a key and the public engagement key seems to have been split into "integration" and "science communication." It connects SwafS to the Europe 2020 strategy on sustainable, inclusive, and smart growth and introduces the word "open" that would become central to the programme of Commissioner Moedas. At the same time, the idea of "institutional change" as the main impact of SwafS/RRI is highlighted, and there are traces of the AREA framework in that there is a section on "inclusive and anticipatory governance." In terms of calls, RRI scholars were happy to see the reintroduction of research projects proper – so-called Research and Innovation Actions or RIAs – designated to strengthen the knowledge base for SwafS. Finally, perhaps with the recognition that the remaining policy staff simply could not successfully mainstream RRI into all of Horizon 2020, a separate call was designed to help with that. The project that was funded was the already mentioned NewHoRRIzon, arguably one of the more significant SwafS-funded projects.

In 2017, the EC published its interim evaluation of Horizon 2020, an exercise firmly placed within a framework of conventional intervention logic (see Figure 3.4). It focused on efficiency and effectiveness. In its summary, the Commission wrote:

> Science and innovation are long term and risky endeavours creating impact that can only very partially be captured after such a short period. A monitoring system with indicators to systematically track impact (in particular for societal challenges) is found to be wanting.
>
> (European Commission 2017)

That conventional intervention logic and economist mindsets had a strong footing in the Commission Services would hardly be a surprise to anyone and least of all to the SwafS policy staff. Indeed, as mentioned, already in 2013, the still existing DG RTD unit set out to form an expert group on policy indicators for RRI, supported by Lino Paula. More or less in parallel, a tender was called and awarded

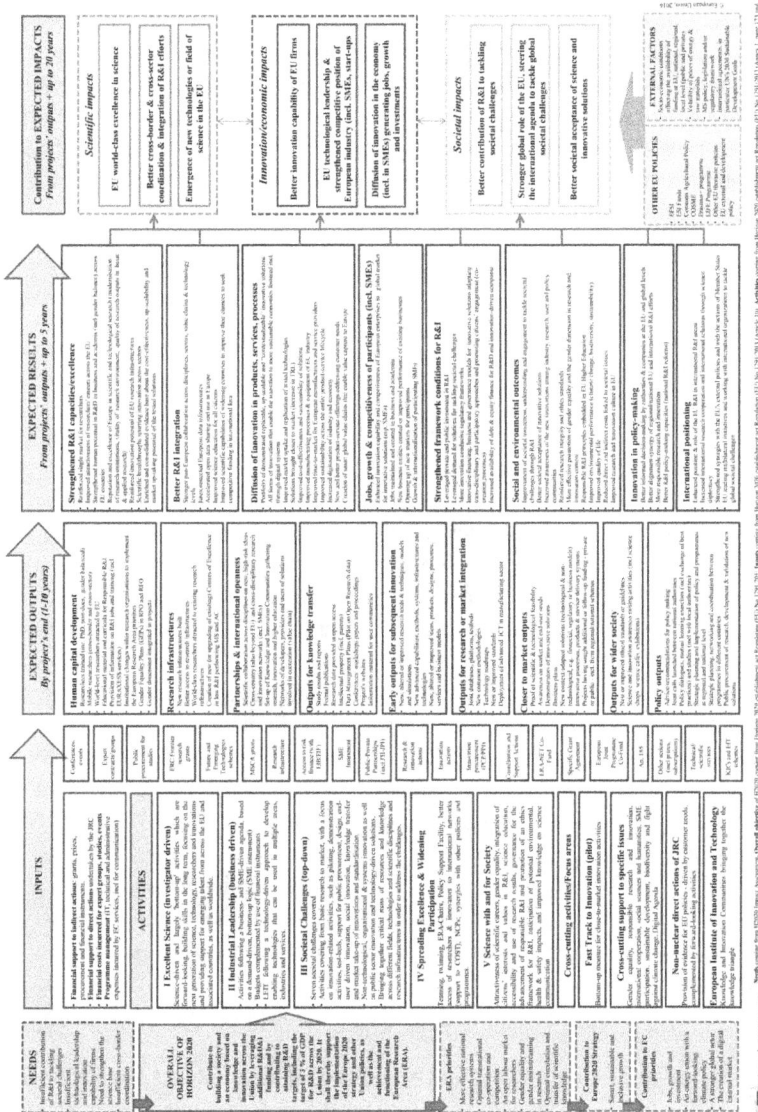

Figure 3.4 Overview graph from the interim evaluation of Horizon 2020.[3]

by what came to be known as the MoRRI project (European Commission 2018), supported by Philip Galiay. MoRRI took a different route than the expert group and produced a set of quantitative indicators that in part were populated with Eurobarometer data and that could be used (although the MoRRI consortium partly warned against that) to compare performance on the member state level.

The 2018–2020 SwafS work programme shows marked influence by the H2020 interim evaluation. The language of the introduction changed. Moedas 3 Os (Open Science, Open Innovation, Open to the World) are now the core justification for SwafS, and in the single mention of RRI on the first page of the introduction, it is connected to citizen science as an instantiation of open science. More important, though, is the prominence of the intervention logic from the interim evaluation. Institutional change has now become the *Key Performance Indicator* for SwafS, and every call is now written to comply with the demand that the expected impact has to be "SMART":

> All applicants should try to detail SMART (Specific, Measurable, Achievable, Realistic, Time-bound) impacts in their proposals, where possible aligned with existing EU or other international objectives. […] The SWAFS Key Performance Indicator (KPI) is related to the number of institutional change actions promoted by the programme. These can take the form of a package of changes across all or several of the five RRI dimensions: Gender (e.g. implementation of Gender Equality Plans), Science education (e.g. introduction of new curricula, new teaching methods, new means of systematically fostering informal learning in non-educational settings), Open access/open data (e.g. introduction of new rules or practices concerning open access and/or open data), Public engagement (e.g. new means of systematically engaging citizens/Civil Society Organisations in research and innovation activities such as through agenda setting, foresight and public outreach), Ethics (e.g. implementation of new rules concerning treatment of research ethics, codes of conduct, ethical reviews).
>
> Several WP18-20 topics specify indicators which applicants should work towards, notably from the Sustainable Development Goals and from the study Monitoring the Evolution and Benefits of Responsible Research and Innovation (MoRRI)
>
> (European Commission 2020, 7–8)

In terms of translation, this change had a profound impact. Some of the political winds and fashions we have described would not necessarily change much at the shop floor level. With the insistence on conventional intervention logic and SMART, however, a set of new practices were introduced. SwafS and indeed all Horizon 2020 grantees or "beneficiaries" as it was called, got new forms to fill in so that the KPIs could be measured. In SwafS, one literally was asked, on an annual basis, to count the number of institutional changes that had been produced. We have never met any participant in a SwafS project who found this practice meaningful; everybody simply invented something to enter into the tables to somehow fulfil the formal requirement. Worse, project proposals would now have to be written in a way so as to argue that the project would "work towards" MoRRI indicators and would be able to document its achievements to that effect. For example, the work programme included a call for research projects (RIAs) to strengthen the knowledge base on citizen science (SwafS-17-2019). Among the expected impacts, the projects:

> […] should aim to indirectly work towards MoRRI indicators (e.g. SLSE4, PE1, PE2, PE3, PE5, PE6, PE7, PE8, PE9, PE10, OA6) and identified and appropriate Sustainable Development Goals.
>
> (European Commission 2020, 23)

The choice of the MoRRI indicators rather than those of the expert group was not surprising. The expert group argued strongly in favour of a concept of network governance and against a simplistic idea of command-and-control. Indeed, the whole field of scholarship on governance of technoscience would argue and warn against conventional intervention logic and see it as part of the problem. And yet here it was, implicitly all along but now imposed by the higher echelons of DG RTD or if not, at least by means of the governmentality residing in the lower ones.

The MoRRI indicators were perfect in this regard – they were largely numerical, composed by statistical information for which there even might be big data, and allowed for multivariate analysis. Yet, what project consortia struggled with was how to make sense of them on a project level. Considering the quote above, the projects were expected to influence policymakers and thereby "work towards" some or all of the list of ten MoRRI indicators. But what were these indicators? How were they defined? SLSE4 is a count of two quantities, namely the number of member organisations in the European Citizen Science Association and the number of scientific publications concerning "citizen science" (European Commission 2018). PE2 is a mean national score of the frequency of affirmative answers to the following three questions:

- Do you attend public meetings or debates about science and technology?
- Do you sign petitions or join street demonstrations on matters of nuclear power, biotechnology, or the environment?
- Do you participate in the activities of a non-governmental organisation dealing with science and technology-related issues?

And so on and so forth. It is difficult to understand how a research project could influence the state of these indicators or if it should do so. It is not even clear what it means to "work towards" these indicators. The phrase would seem to presuppose that the indicators are signals of desirable states, so that if, say, PE2 is 100% that is better than 50% or 30%. But that assumption does not hold scrutiny when viewed against the definition of the indicators, and it was certainly not supported by the publications of the MoRRI consortium. Furthermore, the MoRRI indicators were largely defined on the national scale, which could be quite unsuitable for monitoring the achievements of projects. The SwafS 2016–2017 work programme included a call for an RIA to further the work of MoRRI and, in colloquial terms, sort out part of this mess. The call text was immensely complicated and only one proposal was submitted and funded, the suitably named SuperMoRRI project.[4]

We do not know if the expected impact formulation had a significant impact on the evaluation and selection of proposals to the SwafS-17-2019 call or on the execution of the funded project (called CS-TRACK). Perhaps the actors along the whole chain from evaluation to execution dismissed the requirement as absurd for a research project. Indeed, there is not a single mention of MoRRI indicators on the cstrack.eu website. For the many coordination and support actions funded by SwafS, however, the situation was different in that their direct justification was to achieve change; indeed, to coordinate and support actions to promote institutional change. We shall pursue this further by narrowing our scope to the policy area of our case study, that of regional or territorial RRI.

Territorial RRI

One of the novelties of the 2018–2020 SwafS work programme was the new strategic orientation on "Building the territorial dimension of SwafS partnerships." It was explained with one paragraph of text that mainly listed various categories of actors in a society (universities, authorities, businesses, civil society organisations) and the request or prediction that they will or should work together in "different territorial contexts," adding that "territories are understood as geographical areas sharing common features" (European Commission 2020, 35). It was not the best display of proficiency in the art and science of geography. However, it contained two calls, SwafS-14 and SwafS-22. The latter would seem to be misplaced, as it had nothing to do with RRI or any of the usual topics but was about supporting research in EU's so-called Outermost Regions (the Canary Islands, the Azores and Madeira, and the various territories of Overseas France that are part of the EU). The intuitive home for that call might have been the designated programme on "Spreading Excellence and Widening Participation" that had its own 816 million euros budget; however, since that programme was presented with an explicitly colonial flavour of helping "countries lagging behind"[5] the embarrassment of supporting France, Portugal, and Spain under it might have been intolerable. The first one, however, was the real call of the strategic orientation, and it was opened each of the three years 2018, 2019, and

2020 and funded in all 11 projects with approximately 2 million euros each. The fact that nine of them had RRI in their name (CHERRIES, DigiTeRRI, RRI-LEADERS, RRI2SCALE, SeeRRI, TeRRIFICA, TeRRItoria, TetRRIS, and WBC-RRI.NET) says something about the period and possibly its sense of humour. The two others were called RIPEET and TRANSFORM; this book is about the latter.

The SwafS-14 call text did show insight into regional innovation and development. It argued for the territorial scale (territories being explicitly defined broadly to denote a geographical scale at choice):

> Territories have a specific advantage to address the complexity of the challenges set by the interplay between science and society. Indeed local actors have an intimate knowledge of the physical territorial setting, and local ecology, i.e. the status quo of the complex relationships between cultural, social, economic and political actors, of the local dynamics, history, expectations and requirements as well as specific concerns.
>
> (European Commission 2020, 35)

The idea of the call was to utilise this advantage to introduce RRI ideas and practices in quadruple helix R&I interactions at the territorial scale and work for institutional change also by involving local or regional authorities. Furthermore, the call explicitly mentioned other EU efforts to strengthen regional innovation, notably the instrument of Smart Specialisation Strategies, or S3 as it is often called. All in all, projects should present a "sequence of actions that open up and transform the R&I ecosystem and governance systems so that they are more open and inclusive" (European Commission 2020, 36). In other words, the call has an open approach to what RRI is and could be, in the sense that the scope of work to introduce RRI principles was very wide. A broad range of examples were mentioned as inspiration, including coastal planning, urban development, and land use planning, signalling that RRI could be a matter of quite different things than sophisticated bio- and nanotechnological technoscience.

While the main body of the call text could have been written by a regional development scholar with an interest in RRI, the Expected Impact section was particularly severe in its intervention logic character:

> Expected Impact: Consortia are expected to elaborate and implement a more open, transparent and democratic R&I system in their defined territories. Consortia are expected to evaluate their activities and provide evidence of societal, democratic, environmental, economic and scientific impacts. Involvement in the project should have a measurable transformative and opening effect on organisations involved, which should be sustainable beyond the lifetime of funding. Consortia are expected to contribute to one or more of the MoRRI indicators (for instance GE1, SLSE1,

SLSE4, PE1, PE2, PE5, PE7, PE8, E1, OA6, GOV2), and to the Sustainable Development Goals (for instance goals 4, 5, 9, 11, 12, 13, 16 or 17).

(European Commission 2020, 37)

Quite a few of the projects that were funded, accordingly contained separate monitoring and evaluation activities, such as our case study and the raison d'être of this book, TRANSFORM. Indeed, the set of projects funded in the 2018 and 2019 calls created a designated online forum together with SuperMoRRI to discuss the particular problems of how to perform the evaluation and not the least, how to deal with the set of MoRRI indicators. Gradually, the approach of evaluative inquiry, as chosen in this book, emerged as a better approach than a perfunctory exercise with more or less meaningless MoRRI indicators (even if the latter also had to be done in project reporting).

A particular aspect of translation can be identified in this regard, which of course is not unique to RRI projects at all but is part of what has been called "projectification" (Torka 2006; Wicher and Frankus 2023). The expectation of providing evidence of impacts within a three-year period constitutes pressure towards project designs and problem definitions that have a short time frame. Indeed, for *impact* to be observed in a three-year project, the time frame of what to do should ideally be shorter, say, one or two years. It calls for a focus on low-hanging fruits.

The TRANSFORM proposal

TRANSFORM was one of the five projects that received funding in the second SwafS-14-call and began its work in 2020. The core idea of the project proposal was to explore three different methodological approaches to RRI at a regional scale – participatory agenda-setting, citizen science, and design thinking – in three regions of Lombardy, Catalonia, and Brussels-Capital-Region, respectively. In addition, the project had a (virtually non-funded) arm in the US, drawing on a collaboration with the Boston Museum of Science. The coordinator of the initiative was Angela Simone at the *Fondazione Giannino Bassetti* in Milan, a foundation that had a long-standing interest in its own conception of responsible innovation that was even inscribed into regional law; indeed, the founder of the Bassetti Foundation, Piero Bassetti, is an influential public figure in Milan who was the first president of Lombardy (see also Chapter 4). Simone developed the proposal in close collaboration with Marzia Mazzonetto who set out to lead the Brussels cluster of the project from the Belgian NGO Be Participation, and Rosa Arias, the founder of the Barcelona-based enterprise *Science for Change,* which specialised in citizen science projects. In each of the clusters, there would be a set of other project participants, including academic partners and regional authorities involved in regional smart specialisation policies. The idea was then to perform pilots and other activities that would embed

RRI practices above all in the regional authorities and thereby embed RRI in regional innovation S3 plans, the so-called RIS3 strategies. We, the University of Bergen, were contacted to develop and lead a separate work package on Monitoring and Evaluation.

The evaluation that the proposal received was close to panegyric. The proposal obtained a score of 14.5 out of 15 possible points and was selected for funding. Several of the statements from the ESR – the Evaluation Summary Report – are noteworthy. After praising the overall concept of the project, the report states "A number of SDG goals – and the relevant RRIs – are central to the proposal, i.e. endure gender dimension, science education, open access/ open data, public engagement, ethics and environmental sustainability" and furthermore, under the impact criterion: "The proposal will lead to concrete and measurable impacts in transforming the R&I ecosystem and governance systems in the project regions [...] The expected impacts of the project contribute well to several MoRRI indicators and to the Sustainable Development Goals. The impacts are formulated with well-defined measures, which are realistic and achievable."

One should be careful not to overstate the importance of these comments. Taken at face value, they could be interpreted as evidence that the proposal indeed was selected because it aligned itself with the policy discourse of SDGs and MoRRI indicators and because it adopted a style of projectification that aimed for measurable and short-term impacts. If so, this would be a case of a dramatic process of translation from RRI as conceived by von Schomberg and his like, all the way to a command-and-control operation with the RRI keys as the variables and the MoRRI indicators as the benchmarks. There is little reason to take such statements wholly at face value, however. Rather, from research on research evaluation it is known that reviewers normally come to a global conclusion on a proposal – individually and collectively – and then they are required to write an assessment report that not necessarily reflects their most deeply held reasons (Brunet and Müller 2022). Rather, the assessment report has its own genre requirements that are set by the call and the organisational culture of the research funding organisation. Those who have had their own experience from the EU review panel may recognise that the organisational culture is loaded with norms about what an ESR – an evaluation summary report should look like and what kind of statements it should have. Still, however, even if some of the elements of the process are perfunctory and ritualised, they have to be there, and there has to be a substrate for them. As submitters of the proposal, we did not risk leaving out, say, the MoRRI indicators however little we believed in them. And it is not obvious that the panel would have taken the risk to give such praise and such a high score if these elements had not been present. In sum, the translation takes place, albeit not necessarily to the degree that a naïve reading might indicate. It is there and it is not there. The proposal is selected, it is converted to a grant agreement with a set of contractual obligations, and what is written within the discourse of the call, is part of those obligations. At the same time, there is still room for the proposers to invest their creative ideas and moral convictions into the proposal, for the evaluators to be enthused by them, and the operational phase of the project to be guided by them. In the subsequent chapters, we try to show evidence thereof.

Summing up

At the beginning of this chapter we recalled Pope Francis' warning about the one-dimensional technocratic paradigm. The concept of RRI and the intellectual tradition it emerged from were critical responses to that paradigm and attempts at changing it. The story of RRI and SwafS is a story about how a quite marginal voice in European policy, critical towards hegemonic views on the unambiguous blessings of science, innovation, and growth, managed to climb to the top and win a battle by mobilising the European Parliament against the EC. At the same time, it is also a story about how the Empire strikes back. RRI was never mainstreamed in Horizon 2020, and the policy area in DG RTD was severely diminished. The SwafS programme itself became increasingly disciplined and domesticated into the frame of conventional intervention logic, in particular after the interim evaluation of Horizon 2020 in 2017. We also noted that the RRI community, what Shanley (Shanley, 2022) called a scientific-intellectual movement, perhaps even an epistemic community in Haas' (1992) sense, adopted an orientation towards expecting and assessing impact and developing indicators.

Still, SwafS took place and it gave rise to hundreds of projects all across Europe with thousands of individuals involved. Even if the technocratic paradigm ruled in Brussels, at least in frontstage discourse, creative translations of RRI could take place everywhere else, and surely also in Brussels, backstage, in the heroic efforts from the few policy staff left in DG RTD who fought tooth and claw to keep as much as the original sense of RRI present. As mentioned above, there is no scarcity of papers and book chapters that complain that the ideal implementation of RRI never took place. But how could it, given the powers of anti-RRI? (Strand 2019)

Around 2014, as it was becoming clear that there would be no EC Recommendation on RRI and no real mainstreaming of RRI into Horizon 2020, some policy officers advised us researchers to focus efforts on smaller geographical scales, in the member states and in region or local place. This was in effect what SwafS to a large degree also tried to facilitate. Their reason was that at the moment, no more could be achieved at the centre; we had to continue the quest in the many peripheries. That view resonated strongly with the theoretical perspectives to which RRI belonged – it was a call to extend the networks, to let the rhizomes grow, or, in the words of Deleuze and Guattarí, to *become* new minorities and thereby create new forms of agency.

At the end of this chapter, we accordingly prepare to change the focus to get a view of what happens locally and what virtues it holds even if it does not match the blueprint of RRI's imperial aspirations. It seems appropriate to return to the encyclical letter:

Attempts to resolve all problems through uniform regulations or technical interventions can lead to overlooking the complexities of local problems which demand the active participation of all members of the community. New processes taking shape cannot always fit into frameworks imported from outside; they need to be based in the local culture itself. As life and the world are dynamic realities, so our care for the world must also be flexible and dynamic.

(Francis 2015, Article 144)

Notes

1 The personal observations in this chapter are mainly due to one coauthor, Roger Strand.
2 AREA was an acronym for Anticipate, Reflect, Engage, and Act as the four dimensions of how to "do" RRI/RI (Stilgoe et al. 2013).
3 See https://research-and-innovation.ec.europa.eu/knowledge-publications-tools-and-data/publications/all-publications/interim-evaluation-horizon-2020-book-version_en. Last accessed April 25, 2023.
4 https://super-morri.eu/. Accessed March 30, 2023.
5 https://research-and-innovation.ec.europa.eu/funding/funding-opportunities/funding-programmes-and-open-calls/horizon-europe/widening-participation-and-spreading-excellence_en

References

Åm, Heidrun. 2019. "Limits of Decentered Governance in Science-Society Policies." *Journal of Responsible Innovation* 6 (2): 163–78. https://doi.org/10.1080/23299460.2019.1605483.

Balmer, Andrew S, and Kate J Bulpin. 2013. "Left to Their Own Devices: Post-ELSI, Ethical Equipment and the International Genetically Engineered Machine (IGEM) Competition." *BioSocieties* 8: 311–35. https://doi.org/10.1057/biosoc.2013.13.

Balmer, Andrew S, Jane Calvert, Claire Marris, Susan Molyneux-Hodgson, Emma Frow, Matthew Kearnes, Kate Bulpin, Pablo Schyfter, Adrian Mackenzie, and Paul Martin. 2016. "Five Rules of Thumb for Post-ELSI Interdisciplinary Collaborations." *Journal of Responsible Innovation* 3 (1): 73–80. https://doi.org/10.1080/23299460.2016.1177867.

Barben, Daniel, Erik Fisher, Cynthia Selin, and David H. Guston. 2007. "Anticipatory Governance of Nanotechnology: Foresight, Engagement, and Integration." In *The Handbook of Science and Technology Studies*, edited by Edward J. Hackett, Olga Amsterdamska, Michael Lynch and Judy Wajcman, 3rd ed., 979–1000. Cambridge, Massachusetts, London: MIT Press.

Brunet, Lucas, and Ruth Müller. 2022. "Making the Cut: How Panel Reviewers Use Evaluation Devices to Select Applications at the European Research Council." *Research Evaluation* 31: 486–97. https://doi.org/10.1093/reseval/rvac040.

Callon, M., and Bruno Latour. 2006. "Le grand Léviathan s'apprivoise-t-il?" In *Sociologie de la traduction. Textes fondateurs*, edited by Madelaine Akrich, Michel Callon and Bruno Latour, 11–32. Transvalor: Presses des mines.

Castro e Silva, Manuela, and Aurora A. C. Teixeira. 2011. "A Bibliometric Account of the Evolution of EE in the Last Two Decades: Is Ecological Economics (Becoming) a Post-Normal Science?" *Ecological Economics* 70 (5): 849–62. https://doi.org/10.1016/J.ECOLECON.2010.11.016.

Daimer, Stephanie, Hendrik Berghäuser, and Ralf Lindner. 2023. "The Institutionalisation of a New Paradigm at Policy Level." In *Putting Responsible Research and Innovation into Practice: A Multi-Stakeholder Approach*, edited by Vincent Blok, 35–56. Dordrecht: Springer.

Davies, S., M. B. Kearnes, and M. Macnaghten. 2010. "Nanotechnology and Public Engagement: A New Kind of (Social) Science?" In *Nano meets macro : social perspectives on nanoscale sciences and technologies*, edited by K. Kjølberg and F. Wickson, 405–423. Singapore: Pan Stanford.

Delgado, Ana, and Heidrun Åm. 2018. "Experiments in Interdisciplinarity: Responsible Research and Innovation and the Public Good." *PLoS Biology* 16 (3). https://doi.org/10.1371/journal.pbio.2003921.

European Commission. 2020. "Horizon 2020 – Work Programme 2018–2020: 16. Science with and for Society." Publications Office.

———, Directorate-General for Research and Innovation. 2008. "Challenging Europe's Research: Rationales for the European Research Area (ERA): Report of the ERA Expert Group." Publications Office. https://data.europa.eu/doi/10.2777/39044.

———. 2012. "Responsible Research and Innovation: Europe's Ability to Respond to Societal Challenges." Publications Office. https://data.europa.eu/doi/10.2777/11739.

———. 2013a. "Nanotechnology: The Invisible Giant Tackling Europe's Future Challenges." Publications Office. https://data.europa.eu/doi/10.2777/62323.

———. 2013b. "Options for Strengthening Responsible Research and Innovation: Report of the Expert Group on the State of Art in Europe on Responsible Research and Innovation." Publications Office. https://data.europa.eu/doi/10.2777/46253.

———. 2015. "Indicators for Promoting and Monitoring Responsible Research and Innovation: Report from the Expert Group on Policy Indicators for Responsible Research and Innovation." Publications Office. https://data.europa.eu/doi/10.2777/9742.

———. 2017. "Interim Evaluation of Horizon 2020: Commission Staff Working Document." Publications Office. https://data.europa.eu/doi/10.2777/220768.

———, C. Spaini, V. Peter, C. Bloch, et al. 2018. "Monitoring the Evolution and Benefits of Responsible Research and Innovation: The Evolution of Responsible Research and Innovation: The Indicators Report." Publications Office. https://data.europa.eu/doi/10.2777/728232.

Fisher, Erik. 2005. "Lessons Learned from the Ethical, Legal and Social Implications Program (ELSI): Planning Societal Implications Research for the National Nanotechnology Program." *Technology in Society* 27 (3): 321–28.

Francis. 2015. "Laudatio Si' Of the Holy Father Francis: On Care for our Common Home." https://www.vatican.va/content/francesco/en/encyclicals/documents/papa-francesco_ 20150524_enciclica-laudato-si.html.

Funtowicz, Silvio O., and Jerome R. Ravetz. 1993 "Science for the Post-Normal Age." *Futures* 25 (7): 739–55. https://doi.org/10.1016/0016-3287(93)90022-L.

Galiay, Philippe. 2011. "Situation in Europe and the World: A Code of Conduct for Responsible European Research in Nanoscience and Nanotechnology." In *Nanoethics and Nanotoxicology*, edited by Philippe Houdy, Marcel Lahmani and Francelyne Marano, 497–509. Dordrecht: Springer.

Glerup, Cecilie, Sarah R. Davies, and Maja Horst. 2017 "'Nothing Really Responsible Goes On Here': Scientists' Experience and Practice of Responsibility." *Journal of Responsible Innovation* 4 (3): 319–36. https://doi.org/10.1080/23299460.2017.1378462.

Griessler, Erich, and Vincent Blok. 2023. "Conclusion: Implementation of Responsible Research and Innovation by Social Labs. Lessons from the Micro- Meso- and Macro Perspective." In *Putting Responsible Research and Innovation into Practice: A Multi-Stakeholder Approach*, edited by Vincent Blok, 273–84. Dordrecht: Springer.

Griessler, Erich, Robert Braun, Magdalena Wicher, and Merve Yorulmaz. 2023. "The Drama of Responsible Research and Innovation: The Ups and Downs of a Policy Concept." In *Putting Responsible Research and Innovation into Practice: A Multi-Stakeholder Approach*, edited by Vincent Blok, 11–34. Dordrecht: Springer.

Haas, Peter M. 1992 "Introduction: Epistemic Communities and International Policy Coordination." *International Organization* 46 (1): 1–35. https://doi.org/10.1017/S0020818300001442.

Hilgartner, Stephen. 1997 "The Sokal Affair in Context." *Technology, & Human Values* 22 (4): 506–22. https://doi.org/10.1177/016224399702200404.

Kjølberg, Kamilla Lein, and Roger Strand. 2011 "Conversations About Responsible Nanoresearch." *Nanoethics* 5 (1): 99–113. https://doi.org/10.1007/s11569-011-0114-2.

Loeber, Anne, Michael J. Bernstein, and Mika Nieminen. 2023. "Implementing Responsible Research and Innovation: From New Public Management to New Public Governance." In *Putting Responsible Research and Innovation into Practice: A Multi-Stakeholder Approach*, edited by Vincent Blok, 211–28. Dordrecht: Springer.

Macq, Hadrien, Élise Tancoigne, and Bruno J. Strasser. 2020. "From Deliberation to Production: Public Participation in Science and Technology Policies of the European Commission (1998–2019)." *Minerva* 58 (4): 489–512. https://doi.org/10.1007/s11024-020-09405-6.

Nordmann, Alfred. 2007. "If and Then: A Critique of Speculative Nanoethics." *Nano Ethics* 1 (1): 31–46. https://doi.org/10.1007/s11569-007-0007-6.

Oudheusden, Michiel van, Nathan Charlier, Benedikt Rosskampa, and Pierre Delvennea. 2015. "Broadening, Deepening, and Governing Innovation: Flemish Technology Assessment in Historical and Socio-Political Perspective." *Research Policy* 44: 1877–86. https://doi.org/10.1016/j.respol.2015.06.010.

Owen, Richard, Phil Macnaghten, and Jack Stilgoe. 2012. "Responsible Research and Innovation: From Science in Society to Science for Society, with Society." *Science and Public Policy* 39 (6): 751–60. https://doi.org/10.1093/scipol/scs093.

Owen, Richard, René von Schomberg, and Phil Macnaghten. 2021. "An Unfinished Journey? Reflections on a Decade of Responsible Research and Innovation." *Journal* of *Responsible Innovation* 8 (2): 217–33. https://doi.org/10.1080/23299460.2021. 1948789.

Pickering, Andrew. 1992. *Science as Practice and Culture*. Chicago: The University of Chicago Press.

Ribeiro, Barbara, Lars Bengtsson, Paul Benneworth, Susanne Bührer, Elena Castro-Martínez, Meiken Hansen, Katharina Jarmai, et al. 2018. "Introducing the Dilemma of Societal Alignment for Inclusive and Responsible Research and Innovation." *Journal of Responsible Innovation* 5 (3): 316–31. Routledge. https://doi.org/10.1080/23299460. 2018.1495033.

Rip, Arie. 2016 "The Clothes of the Emperor. An Essay on RRI in and around Brussels." *Journal of Responsible Innovation* 3 (3): 290–304. https://doi.org/10.1080/23299460. 2016.1255701.

Shanley, Danielle. (2022). Making responsibility matter: the emergence of responsible innovation as an intellectual movement. Doctoral Thesis, Maastricht University. https://doi.org/10.26481/dis.20221208ds.

Shapin, Steven. 1998. *The Scientific Revolution*. Chicago: The University of Chicago Press.

Stilgoe, Jack, Richard Owen, and Phil Macnaghten. 2013. "Developing a Framework for Responsible Innovation." *Research Policy* 42: 1568–80. https://doi.org/10.1016/j. respol.2013.05.008.

Strand, Roger. 2019. "Striving for Reflexive Science." *Fteval – Journal for Research and Technology Policy Evaluation* 48: 56–61. https://doi.org/10.22163/fteval.2019.368.

_____, and Jack Spaapen. 2021. "Locomotive Breath? Post Festum Reflections on the EC Expert Group on Policy Indicators for Responsible Research and Innovation." In: *Assessment of Responsible Innovation. Methods and Practices*, edited by Emad Yaghmaei and Ibo van de Poel, 42–59. Routledge. https://doi.org/10.4324/9780429298998.

Torka, Marc. 2006. "Die Projektförmigkeit der Forschung." *Die Hochschule* 1 (2006): 63–83. https://doi.org/10.25656/01:16427.

von Schomberg, René. 2011. "Towards Responsible Research and Innovation in the Information and Communication Technologies and Security Technologies Fields." Publications Office. https://data.europa.eu/doi/10.2777/58723.

_____. 2007. From the Ethics of Technology towards an Ethics of Knowledge Policy & Knowledge Assessment. https://doi.org/10.2139/ssrn.2436380.

_____. 2013. "A Vision of Responsible Research and Innovation." In: *Responsible Innovation: Managing the Responsible Emergence of Science and Innovation in Society*, edited by Owen, Richard, John Bessant and Maggy Heintz, 51–74. John Wiley & Sons. https://doi.org/10.1002/9781118551424.ch3.

Wicher, Magdalena, and Elisabeth Frankus. 2023. "Research governance for change: funding project-based measures in the field of responsible research and innovation (RRI) and their potential for organisational learning", *The Learning Organization,* Vol. ahead-of-print No. ahead-of-print. https://doi.org/10.1108/TLO-10-2021-0119.

Yaghmaei, Emad, and Ibo van de Poel. 2020. "Introduction." In *Assessment of Responsible Innovation: Methods and Practices*, edited by Emad Yaghmaei and Ibo van de Poel, 1–8. Routledge. https://doi.org/10.4324/9780429298998.

4 Institutionalising deliberative democracy in Lombardy

Stories of responsibility and chandeliers

It is December 2021, and we are about to meet our Lombardian project partners in-person for the first time after nearly two years of online project meetings. When we arrive at Milano Centrale – the main train station – we realise that the city is already in full-on Christmas mode. The streets and the myriad shops are decorated with beautiful festive lights. Fondazione Giannino Bassetti (FGB), where we will meet our colleagues, is in walking distance, so we decide to use the opportunity for a little walk through the city. We pass old buildings until we arrive at Porta Venezia, also fully decorated with Christmas lights. We cross the street, trying not to become a casualty of Italian traffic, and are now almost there.

Arriving at the offices of FGB we are immediately impressed with the ceiling-high wooden bookshelves that cover the walls of the meeting room and are full of volumes on science, innovation, and society. We are offered an espresso and get a little tour through the offices. The rooms of the Fondazione are clearly designed for lectures and debates. We are especially in awe of a huge wooden table and accompanying chandelier in one of the rooms, which look like as if the building was built around them.

The building and FGB offices embody the rich history and FGB and also its standing within the Lombardian research and innovation (R&I) ecosystem as an arbiter of dialogue and debate of issues of responsibility (see Figures 4.1–4.3). The particular translation of responsible research and innovation (RRI) in the TRANSFORM project to be unpacked in this chapter can only be understood properly when taking this historical dimension into account.

It is in this setting that we talk with our colleagues from the Bassetti Foundation about the work that they are currently doing as a part of the TRANSFORM project. We also talk about the work that led up to this and about the work that they hope will follow once this project has ended. We discuss how they think about engagement, participation, and deliberative democracy more broadly as well as about their aims of doing this work. This meant talking

DOI: 10.4324/9781003371229-4

Figure 4.1 Fondazione Giannino Bassetti premises. Andreina and Piero Bassetti in conversation with Roberto Maroni and Gianfelice Rocca. Picture taken by Tommaso Correale Santacroce from Fondazione Bassetti.

Figure 4.2 Fondazione Giannino Bassetti premises – library and lecture room. Picture taken by Tommaso Correale Santacroce from Fondazione Bassetti.

about designing and organising research agenda-setting processes, citizen assemblies, and crucially also about what the members of this cluster refer to as "preparatory stage." This term describes all the nurturing and care work that is necessary to create the conditions for engagement activities that actually matter in regional policy- and decision-making processes. This is work that often goes unnoticed and is considered marginal. In the story we are going to tell about the activities of the Lombardy cluster, we hope to make space for appreciating this type of project work. Before we get to that, it is first important to understand the basic idea of what the Lombardy cluster of the TRANSFORM project wanted to achieve.

Introducing RRI into Lombardy's R&I strategies

In the first couple of chapters of this book we described the basic background of the work done by the various clusters in Lombardy, Catalonia, and the Brussels Capital Region which is provided by the TRANSFORM project. This being a project funded by Horizon 2020 and within the SwafS-14 call more precisely has the overall aim to foster RRI in different European regions. It is, however, not only about promoting RRI for the sake of it. It is about applying RRI ideas, principles, and approaches to improve regional R&I strategies, smart specialisation strategies. In Lombardy, this meant working towards contributions to the Three-Year Strategic Plan for Research, Innovation, and Technology Transfer (PST). This strategy provides the framework for the regional R&I governance for the next 3 years.

At the same time also the regional smart specialisation strategy – the Lombardy S3 – plays a crucial role in the activities of this cluster.

The grant agreement therefore describes a set of interrelated objectives in that regard. The Lombardy cluster activities – in project terminology the cluster was referred to as work package 3 with the name "Participatory Research Agenda Setting - Lombardy Region" – aimed to achieve the following objectives:

- Set-up and carry-out a multi-stakeholder engagement process through participatory research agenda-setting approach.
- Include concrete suggestions, visions, and opinions from citizens and local stakeholders in the next Lombardy Region Three Years R&I Strategic Plan, aligned with regional S3.
- Develop a detailed operation plan which describes the whole process and can ensure the replicability of the approach in Lombardy and beyond.
- Hold a local final event to share learning and outcomes with local R&I and S3 stakeholders.
- Foster novel and transparent governance relations within the regional R&I agenda setting (TRANSFORM Grant Agreement).

Reading this list of tasks and aims we observe that these are quite ambitious. Even more so when taking into account the discrepancy between the scale of Lombardy and its R&I ecosystem on the one side, and one single Horizon 2020-funded project on the other (of course it is not actually one single project trying to achieve these aims, but more on that later).

Lombardy has about 10 million inhabitants, which corresponds to roughly 16% of the Italian population. It produces a GDP of about €400 billion amounting to roughly a fifth of the overall Italian GDP.[1] Lombardy is therefore considered to be one of the richest regions in Europe and presents itself as "an accredited dynamic interlocutor from an economic point of view, and to become one of the most industrialised areas in Europe."[2] Furthermore, Lombardy's higher education sector is highly developed consisting of 13 universities[3] and numerous university spin-offs.

Given this R&I ecosystem, it comes hardly as a surprise that there is an elaborate mix of R&I policy instruments at play in this region. The Smart Specialisation Strategies (S3) are considered to be an ex-ante condition as well as an overall framework for policymaking.

Together with the S3, the Three-Year Strategic Program for Research, Innovation, and Technology Transfer (PST) is an important element in the Lombardian governance framework. This program structures R&I governance at the regional level for 3 year-periods. The current PST for the period from 2021 to 2023 was approved in 2021. It was introduced by law in 2016 in Regional Law no. 29 of the 23rd of November 2016 entitled "Lombardy is Research and Innovation." RRI principles are also enshrined in this law.

There are formal and official impacts of the RRI concept insertion in the Art. 1 of the Regional Law "Lombardy is Research and Innovation." First of all, a novel committee – independent from the Region but supported by the Region as an Advisory Board – has been created to monitor and implement the responsible governance of R&I in Lombardy, called Forum for Research and Innovation. The Forum is composed of ten members who are experts In science and society relationships at large, ranging from social innovation and open data to public engagement and ethics. Members can also come from outside Lombardy or even Italy and are selected after an open call in which candidates must provide a letter of support from a civil society organisation (CSO) to prove their expertise in science and society. Furthermore less visible but also tangible examples of RRI enshrinement is the fact that in many R&I policies in Lombardy, RRI is evoked as a guiding principle or a concrete practice. For instance, in the so-called Lombardia Innovativa initiative to promote and enhance Lombardy R&I by valorising Innovative Models proposed by entrepreneurial excellence in Lombardy. In assessing the models, the Region has also involved the local innovation ecosystem through a community on the regional Open Innovation platform to provide observations on the models so as to support the regional authorities in selecting the best models. One of the criteria to be assessed and commented on by the community is RRI. But in general, RRI permeates the whole approach of designing R&I strategic plans and key policies in the region and experimentation of deeper integration of society is welcome, like the TRANSFORM project and its deliberative exercise. (*Angela Simone, Fondazione Giannino Bassetti*)

The most recent S3 for the period from 2021 to 2027 therefore readily incorporates citizens in the descriptions of challenges that Lombardy wants to address through its policies: "Support industrial transformation towards digital transition and sustainable development to understand the new needs of the citizen as quickly and effectively as possible."[4] The means for achieving this include a particular "'Way of doing' policy under the RRI and open science paradigms."[5]

Within this context, the Lombardy cluster activities consisted of a combination of a participatory research agenda-setting process on just energy transitions and a citizens' jury devoted to the issue of smart mobility. These activities were designed and conducted by FGB and their partners from the regional administration, Lombardy Region – more precisely the General Directorate University, Research and Innovation – and Finlombarda. The goal of these participatory activities as presented on the project website was "to render S3 more inclusive and transparent, ensuring that citizens' voices are heard, and opinions are taken into account in setting up key regional R&I policies (deliberative process)."[6]

Therefore, a first and basic translation is that of RRI into a form of deliberative democracy. What might seem like a trivial observation becomes less trivial when considering that this translation differs from how the other clusters translate RRI, namely as a form of citizen science in the Catalan case and as co-design and co-creation in the Brussels-Capital Region (more about this in the other empirical chapters). Before we will dive deeper into why RRI was translated in this way and how exactly that was done through the work of our Lombardian colleagues, it will therefore be necessary to reflect a bit on the history and conceptual foundations of deliberative democracy.

Participatory agenda setting and citizens' juries in the literature

The citizenry is quite capable of sound deliberation. But deliberative democratization will not just happen.

(Dryzek et al. 2019)

Over the recent years there has been a renewed interest in different forms of deliberative democracy both in academia and beyond. Inter- and supra-national bodies like the OECD or the European Commission have strengthened their efforts to develop and institutionalise novel forms of citizen participation in policy- and decision-making.

In the highly influential report "Innovative Citizen Participation in New Democratic Institutions – Catching the Deliberative Wave" from 2020 (OECD 2020) representative deliberative processes are presented as a way to address the growing complexity of policy-making in the context of the pressing societal challenges of our time. The "ordinary citizen" is having a triumphant return as a central figure in policy- and decision-making. As a collection of case studies, this report understands itself as a guide to policymakers who are interested in establishing and institutionalising forms of deliberation within mechanisms and procedures. Drawing on a long lineage of scholarship on the topic, the OECD describes three main principles of citizen participation and engagement: Deliberation as the careful discussion of certain issues based on information and evidence, representativeness based on random sampling, and impact in the sense of a link to decision-making processes. Following up on this report, the OECD also published evaluation guidelines for deliberative processes (OECD 2021). In these guidelines, five core principles for evaluating initiatives of citizen engagement and participation are spelt out: Independence, transparency, being evidence-based, accessibility, and constructiveness.

The growing importance of engagement and deliberation on a policy level is also illustrated by the launch of the so-called "Competence Centre on Participatory and Deliberative Democracy (CC DEMOS)." The establishment of this Centre is arguably a step towards institutionalising forms of experimenting with different modes of governance. The self-understanding of CC DEMOS is to "support" policymaking. It does so through a number of activities, including capacity building as well as designing and conducting different forms of citizen engagement. The establishment of such an institution within the European Commission might signal a cultural shift towards more participatory forms of policy- and decision-making. A recent report by Alberto Alemanno (2022), commissioned by the European Parliament's Committee on Constitutional Affairs, is further indication of such a shift. In this study, Alemanno argues for the establishment of permanent deliberative mechanisms – focusing on randomly selected European Union (EU) citizens – in EU institutions. Reports like this one are part of a broader discussion on the role deliberative mechanisms can (and should) legitimately have within policy- and decision-making processes at the European level (Jančić 2023).

These attempts of re-shaping European democratic institutions can be traced back (at least) to the White Paper on European governance (COM(2001) 428) where participation was listed as one of five core principles of European governance. The other principles are openness, accountability, effectiveness, and coherence. As understood in this document, participation is envisioned to improve the relevance, effectiveness, and thus overall quality of EU policies through increased opportunities for participation throughout the whole policy cycle.

When we look beyond the policy realm, it becomes clear that the idea of deliberative democracy is in fact quite old – some even trace it back all the way to Aristotle (Dryzek et al. 2019). Different versions of what is currently subsumed under the umbrella terms "deliberative forums" (Niemeyer 2011) or "mini-publics" (Goodin and Dryzek 2006; Grönlund, Bächtiger, and Setälä 2014) actually have quite a long history. They initially emerged in the 1970s as citizen juries and planning cells (Voß and Amelung 2016) before they were further developed at the Jefferson Centre in the 1970s, and then took the form of so-called "consensus conferences" in Denmark through the Danish Board of Technology. The board has a fairly unique position as it was institutionalised as a public body to support the Danish parliament. As such it has been a hub for research and development of public participation approaches and procedures (Voß and Amelung 2016). Courant (2021) distinguishes six generations of deliberative mini-publics: The starting point in his genealogy is the French High Council of the Military Function, a mini-public responsible for dealing with different matters regarding soldiers' working conditions. The next generation in his list is the Citizen's Juries and Planning cells methods that were established in the 1970s and used to draft reports in support of policy processes. He then proceeds to the Consensus Conferences of the Danish Board of Technology in the 1980s, and further to the method of Deliberative Polling that was initiated in the 1990s – a method that introduced the difference between "considered" and "raw" opinions (Mansbridge 2010). Before arriving at the Citizens' Assembly approach he mentions the Oregon Citizens' Initiative Review (established in 2010), a panel that produces information on upcoming referendums.

While different approaches to citizen engagement often have characteristics that set them apart from one another, we should nevertheless recognise their difference: They often bring together a representative sample of citizens from a certain constituency, usually via randomised sampling; they are organised around procedures for reasoned debate and deliberation; and finally, they produce recommendations that are expected to impact policy- and decision-making. Usually, these processes are guided by professional facilitators. During the process, citizens are provided with information by experts before they discuss and develop recommendations on the issues at hand. The idea is that the outcome of these deliberations are consensual judgements of recommendations.

The size of such mini-publics can vary substantially: There are citizens' juries which convene groups of 10–25 citizens, then there are initiatives that bring together around 150 people to deliberate, and for the big constitutional initiatives up to 900 citizens were involved. Referring to these bigger and more impactful initiatives scholars are diagnosing (or announcing) a "constitutional turn" (Suiter and Reuchamps 2016) for deliberative democracy.

Research has shown that citizens are willing to participate in meaningful forms of public engagement and that, if the process is well organised, they are perfectly capable of contributing to high-level deliberations. Furthermore, the focus on thinking together can help overcome tendencies of polarisation, as deliberation promotes considered collective judgement (Dryzek et al. 2019). This then relates to one of the most crucial common elements of deliberative mini-publics, brilliantly summarised by Dryzek et al. in their piece in the journal Science:

> The science of deliberative democracy seeks evidence on the capacities of citizens as they engage democratic dialogue, not as they respond as isolated individuals to survey questions (or even as they respond in social psychological experiments that fail to capture key democratic features).
>
> (Dryzek et al. 2019)

Such forms of engagement are about collective reasoning and thus imply a particular vision of what an "opinion" is, in contrast to visions of what an "opinion" is in normal polling. Whereas polling implicitly or explicitly assumes that an opinion is held by an actor and can simply be "extracted" by means of a survey, deliberative formats, on the other hand, pay tribute to the more dynamic nature of opinion formation. Such formats can therefore allow for opinions to shift and change when confronted with reasoned arguments.

There is one important distinction between citizens' juries and other forms of deliberative mini-publics: They can either be "state-supported" or "civil-society-led" (Courant 2021). This distinction points to the fact that it does inevitably make a difference whether citizen's assemblies are organised top-down or if they emerge bottom-up through the initiative of citizens or civil society organisations. Whereas the former start from an official mandate, are well-funded, and thus able to organise deliberations over a longer period of time, the latter enjoy less institutional support and rely on crowd-funding or sponsoring. In fact, because the latter are more attuned to the actual issues citizens face, they are occasionally turned into state-supported citizens' assemblies. An example of this is the Irish Constitutional Convention and subsequent citizens' assemblies.

The distinction between state-supported and civil-society-led forms of deliberative democracy resonates with a distinction that is more common in the field of science and technology studies (STS) – the distinction between "invited" and uninvited" participation (Wynne 2007, 2011). This distinction addresses similar issues to the one we just discussed but accentuates different aspects of it. Whereas Courant points to the benefits of having state-support int terms of mandate and resources, Wynne is more interested in issues of boundary-drawing, inclusion, and exclusion. He argues that being the one "inviting" others and engaging them goes beyond questions of procedure and material resources. It is about creating the conditions under which citizens are engaged and about carving out their room for manoeuvre. This also includes the identities and repertoires for (inter-)action that participants can then draw on. Being invited, then, always entails being invited in a certain capacity to do certain things (and not others). Having the power to define these things before the deliberation even begins, so the argument goes, creates a power asymmetry that even the best and fairest procedure can not compensate for. Wynne asks: "How are implicit boundaries of public agency and involvement thus set and enforced in the very discursive-practical routines which are allowed to define such supposedly inclusive processes?" (Wynne 2007, 104)

Questions like this one resonate with the challenges and concerns deliberative mini-publics are facing. Such concerns include, for example, implicit assumptions about the participants of public deliberations (Michael 2012; Soneryd and Amelung 2016). As stressed by Wynne and others (Goven 2006; Irwin 2006; Lezaun and Soneryd 2007), neither "the public" nor "the citizen" are fixed or neutral categories. They usually come with expectations about who citizens are and how they should ideally behave. Also "publics" are often treated as static entities that exist independently of certain political issues. Research on participation and engagement, however, shows that the initiatives themselves contribute to creating the public they pretend to engage (Felt and Fochler 2010). Importantly, there is a drive towards consensus formation and thus different procedures associated with deliberative democracy show a tendency to repress differences (Horst and Irwin 2010), which is why scholars have referred to them as "machineries for manufacturing consensus" (Felt and Fochler 2010).

Given these insights from the debate on different modes and formats of deliberative democracy, there is a definitive need on the side of organisers and facilitators of engagement initiatives to constantly reflect on the institutionally stabilised assumptions about the roles assigned to the participants. This includes questions about how the choice of certain approaches over others will impact power asymmetries and thus possible outcomes of the deliberations.

So, while there is ample evidence that citizens are indeed perfectly capable of contributing to democratic policy- and decision-making, there are also numerous empirical studies that point to practical as well as systemic barriers that might hinder meaningful integration of citizens into political processes. That's why we chose this particular introductory quote: Citizens are perfectly capable of making meaningful contributions to political processes on many levels and throughout all phases of the policy cycle. The question is whether they are allowed to contribute. Deliberative democracy does not just happen by itself. Our story about the Lombardian pilot activities addresses exactly this issue: What needs to be done to make deliberative democracy appealing to regional administrations and is this possible in the long-run?

Showcasing participatory agenda setting and citizens' juries

As we described above, the Lombardy cluster designed and conducted a participatory research agenda-setting process on just energy transitions and a citizens' jury addressing the issue of smart mobility.

The participatory research agenda-setting process took place from April to May 2021 and consisted of a survey administered both online and via telephone to Lombardian citizens. In addition to this survey, an online workshop on a just energy transition in Lombardy was organised. The idea guiding this procedure was to combine a quantitative approach – collecting data representing the Lombardy population – with a qualitative format. This qualitative format was a workshop that focused on one of the topics addressed in the quantitative survey.

In addition to this participatory agenda-setting process, the Lombardy cluster also organised a citizens' jury on data-driven smart mobility. This citizens' jury consisted of two meetings on non-consecutive Saturdays with 24 participants in June 2022. The aim was to develop recommendations for funding calls on data-driven smart mobility that the Lombardy Region was preparing as part of its "Smart Mobility and AI" strategy from 2020.[7]

This overall approach resonates with much of the literature on deliberative democracy, as well as with early ambitions on the European level to give a "voice" to the citizens. This has been described as a mode of listening (Dobson 2012). The input of citizens is important precisely because the influence of lobbyists might otherwise become too prominent in European policy- and decision-making. This is what statements about *taking into account the citizens' inputs* hint at. It is the concrete practice of *taking into account* that will determine the quality of the activities of this kind of deliberative format.

The survey was designed by FGB in dialogue with Regione Lombardia and Finlombarda and was further administered by an external agency to a representative sample of people living in Lombardy (approximately one thousand people). Furthermore, the sample was calibrated by age, gender, and province of residence.

> The first one was a survey administered to a representative sample on Lombardy population, calibrated by a series of social demographic variables. And that study was the quantitative part. Then we had the qualitative part that was at this deliberative workshop.
>
> (Int_06)[8]

This survey was combined with a deliberative workshop entitled "Just Energy Transition in Lombardy." To our colleagues from the Lombardy cluster, the choice of this topic illustrates the importance of environmental sustainability to the Lombardian citizens. In addition, the issue of energy transition is relevant across different policy- and decision-making scales, from regional and national all the way to the supra-national level of EU policymaking. The recruiting of the citizens as well as support in facilitation was again provided by an external agency. Due to the COVID-19 pandemic, this workshop was organised online. Still, our colleagues managed to bring together 18 Lombardian citizens, who were then divided into three different groups for break-out sessions to discuss the topic. The workshop was held as a 1-day 8-hour event on a Saturday and was structured as follows: First an informative phase with a brief introduction, an informative video, an expert presentation, and – finally – a Q&A session. Then the workshop proceeded with discussions among the participants, with some time set aside for elaboration on the recommendations. The overarching objective of these activities was to *understand the needs of the citizens*:

> In the, in our case in the participatory agenda setting it was very important to understand the needs of the citizens because the yeah, the activities were focused on these and it was very helpful and also the social demographic variables were very important to understand. Because Lombardy is a very, there is a lot of variety so you have like little villages in the mountains and a big city like Milan.
>
> (Int_06)

Citizens are here framed as representatives of a certain region. What is described as "needs" is a placeholder for the socio-economic situatedness of the respondents to the online survey. The needs were pre-determined in the survey, but the respondents had the option to add or describe additional needs. The list of needs included the following topics: The availability of quality food in sufficient quantity, disease prevention, questions of health care in the different regions (timely and appropriate care close to home, treatment or rehabilitation at home whenever possible, innovative care and therapies), culture and art, communication and information accessibility, services for citizens (e.g., registry office, school enrolments, medical reservations), safe mobility (such as pedestrian, cyclist, motorist) and public transport, safety of living conditions, multiculturalism, the protection and inclusion of minorities and vulnerable individuals (e.g., the disabled), violence against women, work safety and income.

The second step – a deliberative workshop – had to be conducted online due to COVID-19-related restrictions and was framed as the "qualitative" addition to the quantitative consultation. The objective of this workshop was to collaboratively work on recommendations for work towards a just energy transition. The outcomes of this process were introduced into the most recent Three-Years Strategic Program for Research, Innovation, and Technology Transfer (PST) as *needs* of the population.

Building on these activities and confident because of its success, the Lombardy cluster aimed to establish more ambitious and long-term forms of deliberative democracy. While the initial plan was to establish a citizens' assembly related to Lombardian innovation policies on AI, this was later changed into the format of a citizens' jury. In June 2022, the Lombardy cluster organised such a citizens' jury on data-driven smart mobility. The citizens' jury was organised as two meetings on non-consecutive Saturdays with 24 participants. Again, the participants were selected by an external agency, which also assisted in facilitating the process. The first day was devoted to providing information about smart mobility and its challenges to the citizens. This was the so-called *informative stage*. The inputs given by the experts focused on issues such as responsibility and mobility, privacy, open data, mobility, and gender. After receiving these inputs, the participants were divided into different break-out sessions where they collaboratively developed questions for the experts in a subsequent Q&A. During the second day of the citizens' jury, recommendations were developed and then at the end of the meeting presented to representatives of the Lombardy Region. According to our colleagues from Lombardy, these recommendations are informing the design of a call for funding named "Smart Mobility Data Driven," which itself is part of the Lombardian Smart Mobility and AI strategy from 2020.

Both pilot activities had a clear objective as well as a plan on how to impact regional policymaking. There is, however, an underlying second-order objective to these activities. The aim was not only to better understand the needs of the Lombardian citizens and subsequently to provide better – in the sense of more tailored – policies. In addition, our colleagues from Lombardy wanted to collaboratively design and conduct a methodologically sound, representative *consultation* as a showcase for their partners within the Lombardy Region.

> So, the first was to really to think about representativeness of the citizens to have this broad survey with sample of representation of a population in Lombardy. So, to show the Region that that means to consult your population in a, from a strong methodological point of view (…)
>
> (Int_05)

Such a showcase needed to be provided *from a strong methodological point of view* according to our colleagues, so that they could use the pilot activities to build capacity within the Region, and also to help convince actors who were still hesitant. In doing so, FGB could rely on the help of actors from the regional administration who were already willing to participate in this project. This is a crucial point to which we will return later in the comparative parts of this book.

Before we unpack the activities of the Lombardy cluster, we want to talk about the institutional set-up of this cluster. The situation in this cluster is rather interesting and allows our colleagues to achieve what the other clusters could not. However, their approach also comes with its own set of challenges and risks. So, who or what is the Bassetti Foundation – or FGB – and what is their place within the Lombardian R&I ecosystem?

Fondazione Giannino Bassetti and the Lombardian R&I ecosystem

One of the central elements of the translation of RRI into the Lombardian R&I ecosystem is the position of FGB within this ecosystem and especially its relation to the regional administration represented by Regione Lombardia and Finlombarda. Formally, FGB is a Foundation of Participation since 2016, which means that it can collaborate with external *participants*. The *first participant* is the regional government represented by its President within FGB's Board of Participants. FGB's president Piero Bassetti was the first President of the Lombardy Region from 1970 to 1974, and since its inception in 1994, the Foundation has worked together with the Region on issues of responsibility in innovation governance. This shows that there are close ties between FGB and the Lombardy Region, both formal and informal. The mission statement of FGB mentions a version of responsible innovation and states that it aims to "create a new and renewed awareness around the memory of a precedent, a modern and widespread sense of social, civil and political responsibility amongst those who innovate."

This means that there is a well-established and institutionally anchored history of collaboration between the different partners in this cluster. There is a clear distribution of responsibilities and roles among the various organisational actors and a shared understanding of the purpose of their collaboration both in terms of the separate organisations' interests but also when it comes to the aim of developing and nurturing a culture of RRI. In this region, this takes the shape of an R&I governance that increasingly relies on novel forms of deliberative democracy, such as participatory agenda setting, citizens' juries, and at some point in the future – this is at least the ambition and hope of FGB – citizens' assemblies.

More recently, FGB played a consulting role when the Region aimed at strengthening the legal foundation for RRI within the regional governance system. The result of this ambition was that RRI now is a part of the *Legge Regionale 23 novembre 2016, n. 29*, which we already introduced. This law states that to strengthen regional innovation and the competitiveness of the system a culture of RRI needs to be established. This is to be achieved by disseminating and experimenting with innovative methods and processes. In addition, it also provides the legal basis for the *Forum for Research and Innovation* which has an advisory function. FGB is formally recognised as a supporting body to this forum.[9]

What is interesting in this institutional configuration is that long before the acronym RRI became fashionable in the EU policy realm, FGB had already made *responsibility* one of its core pillars. It therefore comes as no surprise that the relation between the different partners in the TRANSFORM Lombardy cluster is built around the idea of responsibility. One of our colleagues put it like this:

> But (laughter) if we say if we say responsible innovation, people have at least an idea okay. If we talk about the general directorate whether in

some works then it is something where people more or less have a clear understanding. Okay. It's a less vague notion. (…) That people who have been in touch with Fondazione Bassetti, normally General Directorate for Research and Innovation, know that Bassetti Foundation is responsible innovation and Fondazione tries to carry on a discourse on what responsibility/what kind of responsibility the regional administration has when it does innovation policy.

(Int_07)

The particular set-up of the Lombardian R&I ecosystem allows for things to happen that wouldn't be possible somewhere else. However, there are also certain challenges that come with relationships like this one. In what follows, we want to unpack this translation and the network with which it co-emerges and then point to those opportunities and risks as they become visible in the work of the TRANSFORM Lombardy cluster.

Participatory agenda setting and citizens' juries as translations of RRI

We have thus far spent some time describing the *what* of the Lombardy cluster's activities. We now want to delve into the *hows* and *whys* of it all. This means unpacking the activities and the ideas and concepts that are guiding them. We will start by looking at what is considered *good* engagement by our colleagues in Lombardy and how this relates to the models of engagement and subject positions that become visible in their accounts of what they are doing. Following that we will describe how the activity of *piloting* or *showcasing* is enacted in the work of this cluster and how the distinction between a *preparatory* and *main* stage of the activities plays a role here.

Good engagement in the Lombardy cluster

One of the most interesting themes in the conversations with our colleagues from the Lombardy cluster was their idea about what constitutes legitimate purposes of RRI work in Lombardy. In that regard, the rationales guiding the activities in this cluster are very clear and transparent. First and foremost, conducting pilot projects following the principles of RRI means working with policy- and decision-makers.

And so, the responsible, all the reflections and activities of Bassetti Foundation really revolve around the concept of science and its interaction with the policy-making environment and the actors and dynamics and activity. So, you can talk about and you can make responsible innovation if you work with people really involved and key in governing innovation, which means not only of course policymakers but the people in charge of decision-making in governance and in the government, in the governments to govern research and innovation. Of course, you can do terrific work also with innovators, with researchers but a relevant, crucial point is working with policymakers and decisionmakers.

(In_05)

What becomes visible in statements like this one is an emphasis on policy-making. RRI is translated as a particular form of innovation governance that is built on a top-down notion of governance. To change how innovation systems work, so the argument goes, you need to start with the people who are responsible for setting the framework conditions in a given innovation ecosystem. Consequentially, policy- and decision-makers are the primary collaborators in working towards more RRI systems.

Not surprisingly then, the way in which the activities are set up in this cluster mainly centres around innovation and its governance. Science and research, while playing a role in the work of FGB more generally, play a minor role in the TRANSFORM project.

While the cluster's participatory activities of course centre around technoscientific innovation, this translation of RRI is not mainly interested in the responsible governance of research institutions. Rather, it clearly focuses on innovation and on the production of innovation strategies. This is not surprising, as one of the main objectives of the project is to influence the development of the Regional Smart Specialisation Strategies (S3) and – in the case of the Lombardy cluster – the Three-Year Strategic Plan for Research, Innovation, and Technological Transfer (PST). The RRI part in this is integrating *citizens needs* better and developing more participatory modes of setting R&I agendas to make sure this can actually happen.

This focus also clearly shapes our colleagues' understanding of what constitutes *good engagement*. Good in our conversations is very often used synonymously with *real* and as such it is demarcated from *fake* participation:

We have a long history for responsible innovation. So, I don't want to have fake participatory process so just to say or to have a deliverable or to tweet on Twitter. Oh, we have made this wonderful workshop collecting recommendations that no one in the world will read and use somehow.

(Int_05)

Statements like this and the distinctions that are drawn point to a very distinct set of challenges and risks associated with designing and conducting engagement activities. These risks that our colleagues highlight have been described in the literature on participation and engagement as a risk that engagement activities are merely "symbolic" (Dryzek et al. 2019) or that they are used as a form of "tokenism" (Arnstein 1969). Used in this way, participation becomes a mere means of window-dressing decisions that have already been made. As such, this distinction between *real* and *fake* points to perceived risks in the work of the Lombardy cluster. It is also a challenge that comes with working alongside administrative or government actors (Völker and Pereira 2023).

Fake here indicates engagement without any commitment to act on what has been discussed. One of the main priorities in the work of the Lombardy cluster is to make sure that the input of the citizens does actually matter in innovation governance processes. This ambition is clearly guiding how the engagement activities are set up and how partners are selected. The aim to make engagements matter is also shaping ideas about the long-term impacts that transpire as a result of the TRANSFORM project. In short, the whole process of the Lombardy cluster is organised according to this overarching principle of having meaningful and real engagements. While such an ambition is commendable, there are of course also some challenges that come with such a strategy. We will return to this point later. Before we do that, however, we want to unpack the different elements of our Lombardian colleagues model of engagement a bit further.

Models of engagement and subject positions

Ideas about what constitutes *good* engagement are expressed in our colleagues' views on the actors who become involved in the deliberations, the roles they should play, and how the outcomes of these activities are supposed to enter the policy- and decision-making processes in Lombardy.

As we described in the beginning of this chapter, a common feature of different versions of deliberative democracy is that usually citizens are put together with experts to deliberate on a certain topic or issue. These deliberations are guided by professional facilitators. The outcomes of such deliberations are then fed back into the policy realm where ideally the suggestions and ideas from the citizens will have a noticeable impact on policy- and decision-making.

This basic model is also applied in the work of the Lombardy cluster and the main actors are virtually the same. In terms of deliberative democracy, this is a kind of "state mandated" initiative but the funding body being the EU. Therefore, the mandate is not as strong in terms of being consequential as compared, for example, to the Irish citizens' assemblies that had concrete legal impacts.[10] This version of RRI is mediated by the practice of *experimenting* or *piloting*, which means that the activities are only partially focused on the issues at hand. Much rather the focus is to conduct a good showcase for the Lombardy Region so they can be convinced about the approach and at the same time learn how to replicate it for other issues.

The overarching objective of these activities was to *understand the needs of the citizens*:

> In our case in the participatory agenda-setting it was very important to understand the needs of the citizens because the activities were focused on these and it was very helpful and also the socio-demographic variables were very important to understand.
>
> (Int_06)

Citizens are engaged as holders of certain *needs* that can be known through different kinds of survey and interview methods. Citizens are thus invited as experts for needs of a region and their lived experience in these regions. They are put together with experts for different technological areas such as big data and AI, open data, privacy or vehicle construction. In addition, an expert on inclusion and gender equity was part of the citizens' jury. In some instances, these experts are also part of regional innovation clusters and have thus knowledge of and stakes in regional innovation policy.

This conceptualisation also ties into a particular theory of change in which these needs of the citizenry must be expressed to be properly understood. Once these needs are understood by the Lombardy Region, innovation strategies that are developed by the regional government together with different innovation clusters (populated with stakeholders from industries and academia) are supposed to address these needs. The relationship between innovation policy and society in this translation of RRI is thus one of providing solutions to problems. The added value as described here is that through this engagement there can be some regional (provinces) specificity to what the Region is doing.

What is a pilot?

The engagement activities of the different TRANSFORM clusters in Lombardy, Catalonia, and Brussels-Capital Region are framed as pilot projects. The different clusters use this term to different extents and also the term *pilot* means different things in the different clusters. In some cases, they mean different things for the partners within single clusters. The ways in which the cluster activities are understood and used differ in terms of the general purpose of the activities, their particular objectives, and also with regard to the perceptions of risks and potentials associated with them. The conduct of such an activity, for example, can take the meaning of designing an activity and *piloting* it in the sense of tinkering and experimenting with it. Indeed, the administrative partners in Lombardy use both terms, *pilot* and *experimentation* (Int_08). The goal then is to fine-tune the approach and methodology. Therefore, it is important to carve out what our colleagues talk about when they use terms like piloting or experimenting with regard to their cluster activities.

In the Lombardy cluster, one of the main goals of the activities is to *showcase* the potential of participatory or deliberative approaches. This is described by our colleagues as showing the partners from the regional administration something *concrete*:

> Transform is now I think is providing (…) let's say a concrete/it's something concrete that shows to [NN] and the rest of the group that responsible innovation is more than a principle."
>
> (Int_07)

Something *concrete* here is framed in terms of methodological choices as this can be seen as a showcase for what it means to apply the *principles* of RRI to governance processes:

> So, considering the timing and the constraints to deliver the PST [the Three-Year Strategic Plan for Research, Innovation, and Technology Transfer; T.V.], we decided to go for an online survey and online workshop to reach a double achievement. So, the first thing was to really think about representativeness of the citizens to have this broad survey with sample of representation of a population in Lombardy. So, to show the Region that that means to consult your population from a strong methodological point of view but also to start to show what means having deliberation, what is the strength of a qualitative exercise with an online workshop.
>
> (Int_05)

What piloting or experimenting thus entails for our Lombardian colleagues is clarified through their emphasis on *showing something concrete*. This approach does not understand the cluster activities in the sense of experimentation as tinkering. It is not a series of trials and errors to find the best solutions to a given problem in a certain context. Rather, piloting as it is described in the statement above clearly points to something different than slowly developing or fine-tuning a certain approach or methodology. The approach is already mature and ready to be presented. Instead, the activities in Lombardy were designed to do two things: Firstly, to showcase what can be done through engagement methods, that is, demonstrate what the added value is. A pilot in this sense is a demonstrator.

In addition, and this is crucial to note here, the activities in the Lombardy cluster are about showcasing diversity, showing that there are different ways to engage the public. The administrative partners from the Lombardy Region are already familiar with polling and conducting surveys. This part was primarily focused on demonstrating how this should be done in a methodologically sound way. The workshop-part of the cluster activity as well as the citizens' assembly mainly focused on introducing something new and proving that this is feasible and that it can indeed be useful and provide an added value.

A second aim the cluster partners from Lombardy kept talking about was that they wanted to give the Region something practical, something they could make use of in the future. *Practical* and *concrete* then take on a different meaning that is more related to capacity building. It is about breaking down the approach in a way so that in can travel – travel from the experts at Fondazione Bassetti to the allies within the different Directorates of Lombardy Region and thus to people within the regional administration who are not yet convinced. This particular translation of RRI as *concrete engagement activities* and blueprints is premised on the relationship between FGB and the Region. These activities and blueprints also help to stabilise this relationship together with a more precise idea of what RRI means in practice.

> We as Lombardy, I think we are strong enough under the legal framework, under the legal point of view to embed RRI in our legal system. (…) And okay, but everybody knows. Okay, so but then what is it?
>
> (Int_07)

RRI is perceived as a well-established idea or principle within innovation governance in Lombardy – thanks to the work of Fondazione Bassetti. In the accounts of our colleagues, this idea is no longer challenged and doesn't need much convincing or explanation. It is even codified in the Lombardian legal system and gets mentioned in innovation strategies such as the PST and the most recent S3. Because RRI is so well-established, the objective of our colleague's work has changed. The activities of the Lombardy cluster are aimed at moving beyond RRI being perceived as an abstract *principle* in the Region. To achieve this objective, methodological guidance and the production of some sort of blueprints are necessary. Here, then, RRI is turned into guidelines for engagement activities that rest on regional legislation.

The activity of piloting and experimenting – conceived in this way – is only possible because it can build on a decade-long history of collaboration as well as on previous work. This is visible, among other things, in institutions like the above-mentioned Legge Regionale with its requirements to initiate RRI activities and report on those activities as well as in the Regional Forum for Research and Innovation.

It is also visible in a particular set-up of the activities that the members of the Lombardy clusters often refer to as a *preparatory stage*. In the next section, we want to show that this stage deserves more attention and also more credit than it usually gets.

The more than preparatory stage in Lombardy

The participatory agenda-setting activities as well as the citizens' jury in the Lombardy cluster are grounded in a broader idea about legitimate purposes and rationales of RRI work. As we described above, first and foremost this means working with decision-makers. This means that there is a strong focus on the governance realm when our colleagues talk about RRI in Lombardy. RRI is understood mainly as a form of innovation governance where policy- and decision-makers are the primary collaborators. In addition to this focus on RRI as collaboration with policy- and decision-makers in innovation governance, the importance of engagement not being *fake* is crucial in the accounts of members from the Lombardy cluster. This stance clearly resonates with work pointing to the risks of window dressing. One of the main ambitions (and priorities) in the work of the Lombardy cluster is to make sure that the input of the citizens does actually matter in the innovation governance work. The question then becomes: How can the Lombardy cluster make sure that input from citizens is used for more than mere window dressing?

In the accounts of our colleagues from this cluster, nurturing and maintaining a good relationship with administrative partners in Lombardy is crucial for this endeavour. A central part of the activities in this cluster then is what can be called *maintenance work*. Maintenance work covers activities that are devoted to building good working relationships with partners in different sectors of the research innovation ecosystem as well as efforts that go into nurturing and curating those relations. This kind of work is usually not formalised in grant agreements and descriptions of tasks and milestones. Nonetheless none of the activities that are covered by such texts – and are thus legitimate objects of performance and impact assessment – would be possible without well-maintained relationships and ecosystems. In the TRANSFORM Lombardy cluster, maintenance involves capacity and awareness-building seminars, collaboratively developing the engagement process and methods, and working towards more visibility of RRI principles on a national level.

In the conversations we had with the members of the Lombardy cluster, maintenance work became visible in the distinction that is often made between a *preparatory stage* and the actual *engagement activities* of the cluster. For this reason, this distinction deserves some more attention here.

In considering the core activities of the Lombardian cluster, the illustration on the project website is a good starting point.

What we see here is the rough outline of the cluster activities of the Lombardy cluster. The main engagement activities were the participatory research agenda-setting process and the citizens' jury. As visualised in this illustration, several smaller activities were organised around these two core activities. Mutual learning, process design, and capacity building are depicted here as actual part of the timeline, whereas some of the other ongoing activities FGB has been doing are delegated into a separate box. So, what we see visualised here is the distinction between what our colleagues referred to as *preparatory stage* in contrast to the *concrete* cluster activities.

So, before the concrete starting of the engagement stage, we had a preparatory stage which was the key to be sure that the public engagement activities were actually actionable from the Region. So, we had a lot of

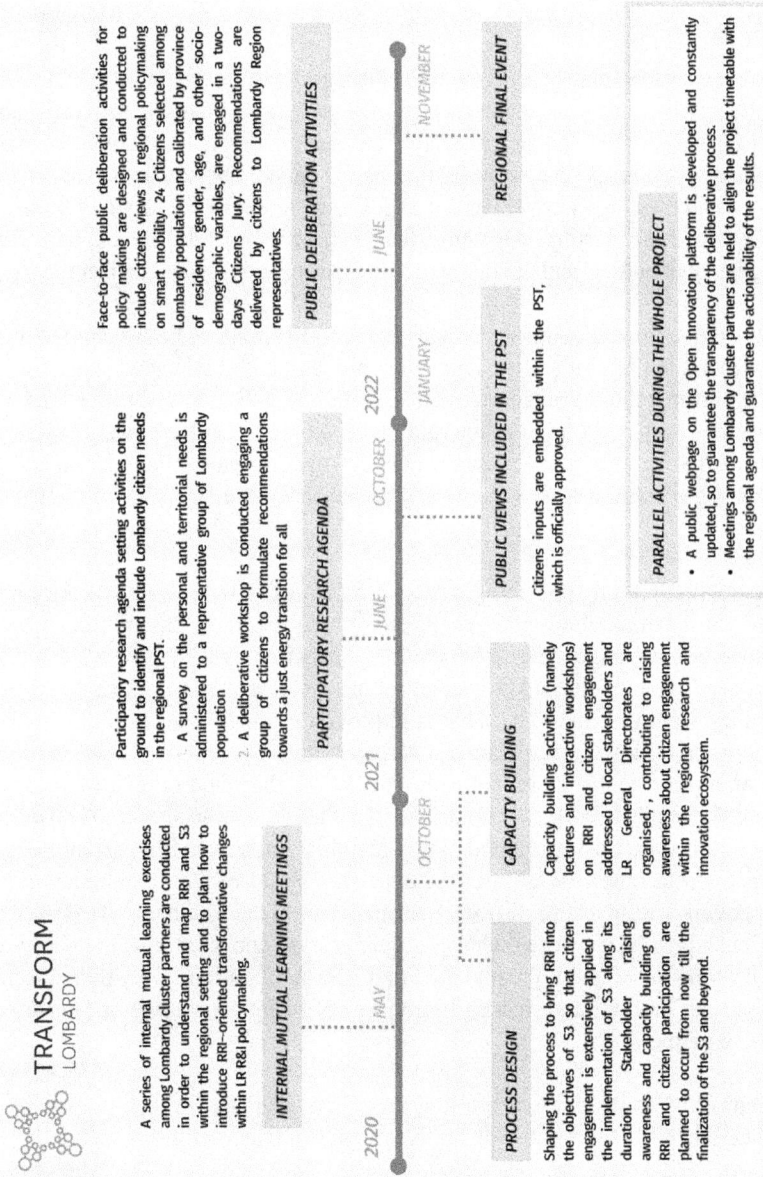

TRANSFORM
LOMBARDY

INTERNAL MUTUAL LEARNING MEETINGS

A series of internal mutual learning exercises among Lombardy cluster partners are conducted in order to understand and map RRI and S3 within the regional setting and to plan how to introduce RRI–oriented transformative changes within LR R&I policymaking.

PROCESS DESIGN

Shaping the process to bring RRI into the objectives of S3 so that citizen engagement is extensively applied in the implementation of S3 along its duration. Stakeholder raising awareness and capacity building on RRI and citizen participation are planned to occur from now till the finalization of the S3 and beyond.

CAPACITY BUILDING

Capacity building activities (namely lectures and interactive workshops) on RRI and citizen engagement addressed to local stakeholders and LR General Directorates are organised, , contributing to raising awareness about citizen engagement within the regional research and innovation ecosystem.

PARTICIPATORY RESEARCH AGENDA

Participatory research agenda setting activities on the ground to identify and include Lombardy citizen needs in the regional PST.

A survey on the personal and territorial needs is administered to a representative group of Lombardy population

2. A deliberative workshop is conducted engaging a group of citizens to formulate recommendations towards a just energy transition for all

PUBLIC VIEWS INCLUDED IN THE PST

Citizens inputs are embedded within the PST, which is officially approved.

PUBLIC DELIBERATION ACTIVITIES

Face-to-face public deliberation activities for policy making are designed and conducted to include citizens views in regional policymaking on smart mobility. 24 Citizens selected among Lombardy population and calibrated by province of residence, gender, age, and other socio-demographic variables, are engaged in a two-days Citizens Jury. Recommendations are delivered by citizens to Lombardy Region representatives.

REGIONAL FINAL EVENT

PARALLEL ACTIVITIES DURING THE WHOLE PROJECT

- A public webpage on the Open Innovation platform is developed and constantly updated, so to guarantee the transparency of the deliberative process.
- Meetings among Lombardy cluster partners are held to align the project timetable with the regional agenda and guarantee the actionability of the results.

2020 · MAY · OCTOBER · 2021 · JUNE · OCTOBER · 2022 · JANUARY · JUNE · NOVEMBER

Figure 4.3 Visualisation of the cluster activities taken from the project website (https://www.transform-project.eu/lombardy/).

mutual learning meetings with [NN] and Finlombarda team. We talked
about S3 and RRI and kind of mutual training let's say and we discussed
a lot about the S3 procedure.

(Int_06)

One of the key aims of the meetings in this preparatory stage was to make sure
that the activities in the engagement stage were *actionable*. To make them actionable
was to make sure that the activities are organised on a topic that is relevant to the
Lombardy Region and that some form of policy impact can be granted. To achieve
this, several meetings were organised to define the purpose and scope of these activities. But these meetings were more than that. They were also used to build trust
and give input on RRI and various approaches to engagement and deliberation.

We had online meetings and we also met one time here at the Foundation
premises but after the, the engagement process to build the next stage. So,
the first meeting in person was one month ago I would say but we knew
them already and I think as we said in some other meetings it was, I think
it was very important because the connection between the Foundation and
the Region was already very strong. So, there was already a lot of trust.

(Int_06)

Accounts like this one show a crucial aspect of how members of the Lombardy cluster think about good engagement with regard to the different stages.
To have good engagement activities, it is necessary that all partners share an
understanding and a basic knowledge about what it is they want to achieve and
how they want to achieve it. What is important to note here regarding maintenance work is how our colleague carefully points out that this work is premised on a pre-established relationship of trust. At the same time, however, this
part of the project work also contributes to the *maintenance and nurturing* of
this relationship. Bassetti Foundation – and this will become important later in
comparison to other TRANSFORM clusters – could rely on an already existing network with the administrative partners within the Lombardy Region. This
relationship existed before the project even started and, crucially, the people in
this network were already convinced that RRI is important. This points to the
necessity of both mutual trust and some sort of *belief* on the side of the administrative partners:

Yeah. And she's really supporting the citizen engagement activities here
(laughter). It's like yeah. She really believes in these activities.

(Int_06)

What is important to note here is that these actors within the regional administration have a double role in the project and for Fondazione Bassetti beyond

the TRANSFORM project. They help set up the cluster's current engagement activities and make sure they are *actionable* and thus *meaningful* for the citizens. In addition to that, they try to convince colleagues within their own ecosystem about the importance of approaches to governance that in one way or another rest on RRI principles, and in doings so hope to enlist them into the group of people who *believe* in this mode of governing. Without these actors within the administration and without the work they were doing before (and possibly after) the actual TRANSFORM project, the engagement activities of the Lombardy cluster would have very likely turned out differently than they did. Indeed, some of the activities might not have been possible to conduct at all.

When we talked about this work with the regional R&I ecosystem and with their administrative partners in particular, the members of the Lombardy cluster described this relationship as dynamic. The aim is to identify actors within the administration who are interested in alternative modes of governing innovation. Once these actors are found, the next step is to build a relationship with them, so that they slowly become allies within the regional administration:

> I have seen some changes in the relationship for example especially with [NN], because I think I met the team from Finlombarda few times before TRANSFORM. (…) And I think that her approach, her attitude towards this kind of engagement activities really changed a lot. Her knowledge also. (…) Now when we talk we are sure that we are talking about the same things and in the past years it was not.
>
> (Int_06)

This is also where it becomes clear that what our colleagues often refer to as the *preparatory* stage is actually more than preparation, in the sense of getting the organisation of engagement activities right. Of course, it is that too: It is about selecting the right topics, finding people who can assist with planning and facilitation, deciding on the concrete approaches and designs, developing a script, etc. However, there is more to this preparatory stage. What we see at play here is *careful maintenance work* that makes the project activities possible. Once these relationships are established and stabilised to a certain degree, these actors are not only willing to support ongoing projects but are further willing to work towards transformations within the organisational cultures of the regional administration. In a sense, they are then able to effectively *disseminate* this kind of thinking and this way of working together, within the different groups and Directorates and departments of Lombardy Region.

> We are trying to work [NN] is the first person in this on pushing to try to disseminate especially within the other departments and within the Region the results and the process of Transform, because she knows quite well that she can be moved to another department for other activities in,

in any moment. So, for us at the end it's important to disseminate beyond the borders of Transform. But the real dissemination activities are within the Region and within the other departments in the three regions involved.

(Int_05)

Our colleagues, however, are keen to stress that there is a certain fragility to this kind of work. When you rely on individual allies within the administrative ecosystem, there is a risk that you need to start this process over again if these actors are either moved to another group or department, or if they themselves decide to move on. Dissemination then is used in a very particular way by our Lombardian colleagues. Dissemination is not just about making people aware of project results, it is also about convincing potential future allies about the added value of engaging citizens in policy- and decision-making. This is where maintenance work and the understanding and practices of piloting meet. Pilots are needed to showcase the dissemination activities, and because the relationships with allies in the Lombardian Region need continuous maintenance and care:

And that's why as I said I think it's very important that we have these outreach communication activities within the Region, with the other civil servants. We also need to plan the public event to share the TRANSFORM results and also this will be very important.

(Int_06)

Parallel to the work on the project activities – which naturally is the core of the TRANSFORM cluster work and therefore also the most visible work done in the project – the cluster members are constantly reflecting on how to best nurture, stabilise, and expand their network. What we see in the accounts of our Lombardian colleagues is an ongoing process of translation in the more classical sense of enrolment and enactment developed by Latour and Callon (Callon, 1986; Latour, 1987).

(...) it's doing things together and talking together and trying to understand each other, the Bassetti Foundation with its own being a third party and at the same time being part of the cultural environment, shaping the public administrative discourse (laughter). (...) And now [NN] and the other people there perceive their job okay and effectiveness of the job. And that the same time (...) working together okay and trying to make something together.

(Int_07)

The overarching objective of these activities then is framed as an attempt to have some influence on the *cultural environment* of R&I governance in Lombardy and also as an attempt to *shape the administrative discourse*. This clearly resonates with one of the core objectives of RRI, which is to have a transformative effect on the various cultures of innovation governance (Strand and Spaapen, 2021).

While the activities outlined above very much focus on regional-level translations of RRI, part of the activities of FGB also target the national level. These activities, on the surface, are not immediately connected to the TRANSFORM-projects work. However, in the context of RRI initiatives and how these initiatives are working to create a long-term shift in innovation cultures, the overall view changes. One example of this is the recent translation of the OECD report "Innovative Citizen Participation and New Democratic Institutions – Catching the Deliberative Wave":

> We decided to translate the OECD report on the Deliberative Wave and I think it was very interesting for us because of course we had the stronger relationship with the OECD but also with the Region because the launch was very important to also present our work at the national level. Because of course the translation was in Italian, so it was not only for regional, Lombardy regional practitioner and for the Lombardy Region ecosystem but it was for the national one. So, it was a way to enlarge our audience and yeah.
>
> (Int_06)

This quote describes a kind of maintenance work that aims to nurture a culture of deliberative democracy beyond the boundaries of the TRANSFORM project and the Lombardy Region. The *hope* is to "make the Italian context more sensitive to this kind of other approaches" (Int_06). Translating this influential OECD report and thus making it broadly accessible to the national Italian R&I ecosystem positions both the concepts and methods related to deliberative democracy – as well as the Bassetti Foundation as an actor representing those ideas – within the Italian innovation ecosystem. This points to a practice of scaling, that is translating RRI from a trans- or supranational setting into the regional Lombardian and national Italian contexts. What is sometimes imagined to happen more or less automatically by inserting funding into a certain system needs constant – and in terms of project assessment often marginalised and invisible – work that involves adapting and making something fit into a certain context.

Real engagement and impact in Lombardy

We have now talked about the translation of RRI in the Lombardy cluster and addressed various elements of this particular version of RRI as deliberative democracy. In the final part of this chapter, we want to draw attention to some of the challenges that come with this particular idea of RRI as deliberative democracy.

In particular, the notion of *real* engagement deserves some attention, the term our colleagues from Lombardy use to describe their work with partners from the regional administration in order to make sure that the outcomes of engagement activities are relevant to the policy- and decision-making processes.

Before we further unpack this process, it is worthwhile to remind ourselves of some of the central ideas of RRI, as they clearly resonate with the idea about real engagement expressed by our colleagues in Lombardy.

> A transparent, interactive process by which societal actors and innovators become mutually responsive to each other with a view to the (ethical) acceptability, sustainability and societal desirability of the innovation process and its marketable products (in order to allow a proper embedding of scientific and technological advances in our society).
>
> (von Schomberg 2011)

In this quote by René von Schomberg, one of the key actors behind the idea of RRI, responsiveness is a key term. Responsiveness is one of the four dimensions of RRI (Stilgoe, Owen, and Macnaghten, 2013), the other ones are anticipation, reflexivity, and inclusion. The way it is presented here, it mainly refers to responsiveness between innovators and other societal actors in order to improve both innovations (as in marketable products) and the innovation process with regard to acceptability, sustainability, and societal desirability. It is clearly the latter FGB is focusing on in their work with the Lombardy Region. Their work of co-creating cluster activities first and foremost aspires to build capacity within the regional administration to make their processes more responsive and to improve the innovation process as such.

Becoming more responsive indicates some form of shift or transformation both in terms of the actors involved but also with regard to the topics to be addressed and how they are framed. This is not a one-way street of course. Becoming responsive to each other means that all actors involved will be affected as they enter a complex dynamic process of shifting power relations. The interesting question then becomes how the different actors are imagined becoming responsive to each other and how decisions are made about the particular challenges that R&I ecosystems need to become responsive to. The story of the Lombardy cluster can be read as a story about the potential – but also the challenges – of attempts to become more responsive, as the very idea of becoming more responsive privileges some versions of RRI while marginalising others.

When the members of the Lombardy cluster talk about their activities, one of the central elements is the process in which the topics to be deliberated on are identified. In this identification process, the choice of topic for the engagement, the issue to be discussed, as well as the timing of the engagement activity in regard to ongoing processes within the Region, played major roles in their considerations.

Here it is important to keep in mind that the collaboration between the Bassetti Foundation and the Lombardy Region is already established and there is a longstanding relationship between the Foundation and the R&I ecosystem in Lombardy. Compared to the Brussel-Capital Region cluster, for example, this makes it considerably less difficult to identify fitting topics. Fitting topics are those that align with certain windows of opportunity. What this means is that topics can only become relevant for deliberation when there is an ongoing policy-process. In this way, so our colleagues argue, it is much more likely that the inputs of the citizens are earnestly considered by the administration. From the perspective of Fondazione Bassetti and the Lombardy Region, there is only *added-value* to be demonstrated if their pilot activities are addressing topics that are currently relevant within the administrative ecosystem.

While this is perfectly rational, there is an important challenge to this approach. Bassetti Foundation needs to carefully navigate the line between focusing on what is relevant and important to the Lombardy Region at a certain point in time while at the same time avoiding the risk of having the process completely dominated by their administrative partners at the cost of the potential for bottom-up problem framings to emerge. The constellation in Lombardy with its focus on *experimenting, piloting and showcasing* makes it very easy for the administrative partners to tell the Bassetti Foundation where RRI makes sense and where it doesn't.

This notion of *becoming responsive* is especially palpable when our colleagues talk about *feasibility* and *actionability*. Quotes like the one below show our colleagues carefully talking about what can and should be done in a certain institutional-organisational setting:

> I said that the strategy for Transform with [NN] and Finlombarda started to say 'Hey guys, we are here, we have to make something concrete but I know there are some constraints in terms of timing and you have to deliver something at the right time and all these kind of things'. But also, for me it's important not to oversell what we can do and the terms of feasibility for me are relevant and key in this process.
>
> (Int_05)

Note how in talking about what should be done in a certain pilot the relationship between the different partners is always a key element. The pilot has to offer something *concrete* that allows them *to deliver*. Both elements are crucial for maintaining the relationship between the Bassetti Foundation and the Region, as FGB needs to be seen as a reliable partner that does not *oversell* and is able to give the Region something they can work with. Doing something concrete is therefore closely tied to the timing of the pilot, and this is where the notion of relevance comes in. Relevance here is crucially not only a thematic but also a temporal category. Conducting an engagement activity means doing something concrete at a certain point in time. In the fast-moving policy world, reports need to be published before certain deadlines, and as a result, input on particular topics is highly time-sensitive. The societal challenges RRI addresses and to which the TRANSFORM RRI pilots need to be responsive are already filtered by the regional innovation-policy processes and thus tend to have a short lifespan. The risk is that if these activities are not mindful of this temporal aspect, the output that is produced might not be relevant, which in turn means wasting the citizen's time in *fake* engagement. This framing of RRI becomes visible when our colleagues talk about the importance of *honesty*:

> So, I preferred to do less but that's something concrete that we can say that we are honest in this. That we have really done what we have promised. Perhaps just very little success but for me, this can be a first step of longer process but it's something that's honest. It's not something that we have say oh we have done a wonderful citizen assembly and at the end we had just thirty citizens in two meetings discussing not relevant things."
>
> (Int_05)

Honesty here means always having the long-term goal of building a culture of responsible innovation governance. To do so, pilot projects must be ambitious – but not too ambitious. Our Lombardian colleagues in that sense prefer small successes instead of big events that yield no results in terms of policy impact.

This becomes even clearer when our colleagues talk about the quality of engagement. The quality is not defined merely by how closely some blueprint or guideline is followed. Rather, this is about follow-up and the *usability* of the *results*:

> But again, we need to start from the policy point of view for the policy question. So, we need to have a process that can be embedded in the shaping of a policy. So, we cannot do citizen participation process for the sake of saying we are involving citizens. So, you need to use the results for something, and I really would like to know what is this something for you. What is in your pipeline of the policies?
>
> (Int_05)

This means a couple of things: Firstly, there is a need for a result, for something presentable. Second, these results need to be timely, meaning they need to tie into a particular policy discussion that is going on at exactly the time when *the results* are ready. Finally, in terms of planning this means of course that the moment of giving input – of becoming relevant – is the starting point for planning a pilot.

Not every societal challenge can therefore be properly addressed in pilots such as the ones designed and conducted by the Lombardy cluster. A societal challenge needs to be set up in a way so that there can be concrete and actionable results to it. If this is possible there needs to be a process to identify or define the exact moment when results can be actionable and thus be *embedded in the shaping of a policy*. However, this requires considerable skill and effort and is premised on an already existing relationship between the different partners (which needs to be constantly nurtured/maintained).

Such a relationship is the condition for being able to understand the *needs* of the regional administration and their innovation governance plans and practices. In our conversations, the cluster members would often stress the importance of understanding what an institution plans and needs and how citizen engagement fits in there. Importantly, these *needs* are part of the translation, part of the ecosystem, and part of becoming *responsive* as one of the core principles of RRI. So, in a sense, these needs are an integral part of *the territory* or the *local* ecosystem:

> So, I think that to answer to all these challenges it was really key the preparatory work that we had at the beginning because it was crucial to have, to build this common language and also to understand the Region's needs. Because sometimes I have the feeling that when you do citizen engagement, citizen engagement practitioners just want to do their work. But of course, you need to be helpful for the institution you're working with. Otherwise, it's just an exercise without a real impact. So, it was key for us to understand their procedures, their plans, their needs, their yeah.
>
> (Int_06)

Again, statements like this one point to the importance of the preparatory work which is framed here as crucial in terms of later impact. Our interviewee makes a distinction between engagement activities that focus merely on doing the engagement work and their approach which focuses also very much on the *institution you're working with*. A *real impact* in this perspective can only be achieved when you work closely with the regional administration. For this kind of work, common project timeframes do not suffice since these relationships need to be built slowly and carefully. Methodology still plays a crucial role, but with a little twist. It is about *showcasing* how things are done and what is the added value to establish trust and prove the legitimacy of the engagement activities.

While this focus on actionability and *doing something concrete* is certainly a valid strategy in implementing RRI and also for working towards a long-term cultural shift within regional administrations, there is a risk to this type of strategy. The power to define and frame what counts as a relevant societal challenge remains with the administration:

> We stay in the some/somehow the design and the implementation. We have these two. We as TRANSFORM, we have this room to be effective in designing and support implementation.
>
> (Int_07)

Part of the maintenance work we learned about in the conversations with our colleagues from the Lombardy cluster has to do with carefully navigating the *room to be effective in*. Questions about what can be done at a certain point in time and where the limits and boundaries are need to be constantly reflected on. Otherwise, there is a risk of damaging the relationship with the Lombardy Region and thus the possibility of convincing others within the administration of this mode of working. This is, as one interviewee describes it, very much about "what the public administration is willing to share" (Int_07).

This also concerns the roles, subject positions, and interaction repertoires that are available to the citizens participating in the engagement activities designed and set up by the Bassetti Foundation and the Region:

> Because the technology roadmap is really, really technical and complex and sometimes I think it can be difficult to establish a real dialogue with citizens on that kind of applications or elements or artificial intelligence. And so, at the end we will have to produce recommendations that are not so/I had the feeling that a process on that topic would finish to produce a recommendation not very useful at the end for the policy. Not actionable.
>
> (Int_05)

In statements like this one about actionability and feasibility our colleagues – mainly representing their administrative partners and their needs – demarcate between issues, policy documents, and questions that are suitable for engagement and those that are not. This points to ongoing demarcation work mainly done by the cluster partners before the citizens come into play. While perfectly understandable as a strategy to curate and nurture relationships within the regional R&I ecosystem and as a way to work towards cultural shifts within the Lombardy Region, the downside is that citizens tend to become an afterthought. This resonates with longstanding concerns about citizen engagement in policymaking, namely that there are power asymmetries when it comes to the issue of who has the power to grant access and who is able to define and frame the topics that are up for debate (Turnhout et al. 2020).

This is a question of what the citizens are actually supposed to contribute to. The broad idea was to organise a citizen's assembly on AI in Lombardy, more precisely the role of AI in the Region's innovation strategy. When discussing the potential contributions of the citizens to a so-called "technology roadmap" on AI, this was deemed to be too technically complex. The perceived danger, then, was that no *real dialogue* could be established and thus the outcome of the engagement would be not useful for the administration. As we described in the beginning of this chapter, the citizens' assembly – while still an objective down the road – developed into a citizens' jury on data-driven smart mobility. This shift arguably resulted in a more actionable engagement activity that could then feed into a call for funding within a smart mobility funding scheme.[11]

In addition, by focusing on the smart mobility topic, our colleagues from the Lombardy cluster could collaborate with a new actor from the administration, which in turn meant they were able to expand their network. The results of the citizens' jury were presented to different units within the Lombardy Region. According to our colleagues, this has been crucial in order to keep the momentum going.

What we see here is that ideas about being *useful* and producing something *concrete* and *actionable* in order to cater to the *needs* of the regional administration can contribute to demarcation practices and thus influence the space for manoeuvre given to citizen participants before the engagement even starts.

This is not to say that this is necessarily a bad practice or wrong in any way. The important point to consider is that the exclusion or inclusion of citizens – and a certain translation of RRI more broadly – is mediated by the need for constant maintenance of the relationship between FGB and their administrative partners. This focus on maintenance is premised on the position of Bassetti Foundation with the regional R&I ecosystem while at the same time aiming to sustain this position. Translation, in the original sense of Callon and Latour, always carries the meaning of gaining "authority to speak or act on behalf of another actor or force" (Callon and Latour 1981, 279). Bassetti Foundation has established itself in the Lombardian R&I ecosystem as an authority in matters of responsibility and RRI. The particular translation of RRI we see in the accounts and practices of our Lombardian colleagues needs to be understood against this background. It is about maintaining a trusting relationship with the Lombardy Region, but it is at the same time about maintaining its position of authority.

There is of course a tension in this position and in the accounts described in this chapter. This tension is one between aiming for actual citizen empowerment and the transfer of ownership on the one side, and the likelihood of producing something that will have an *impact* in the innovation policy world on the other. There is also tension with other meanings of RRI that focus more on shifting power relations and are geared towards *taming* R&I (see Chapter 3 of this book).

Our colleagues in the Lombardy cluster accentuate having impact on policy processes and strategies – which results in RRI policies entering the current S3 for the period from 2021 to 2027 as well as the most recent Three-Year Strategic Programme – and thus enabling *real* engagement. Other clusters translate RRI in different ways and in doing so focus on different aspects of it. We will now move to the Catalan cluster and their translation of RRI as citizen science.

Notes

1 https://ec.europa.eu/eurostat/databrowser/view/nama_10r_2gdp/default/table?lang=en. Accessed February 23, 2023.
2 https://www.en.regione.lombardia.it/wps/portal/site/en-regione-lombardia/business/industry-and-craft-trade. Accessed February 23, 2023.
3 https://www.en.regione.lombardia.it/wps/portal/site/en-regione-lombardia/DettaglioRedazionale/work-and-education/education/university-system-rl. Accessed February 23, 2023.
4 https://www.s3.regione.lombardia.it/wps/portal/site/s3/lombardy-s3-in-brief. Accessed February 24, 2023.
5 Ibid.
6 https://www.transform-project.eu/lombardy/. Accessed November 23, 2022.
7 https://www.openinnovation.regione.lombardia.it/en/homepage/smart-mobility-&-artificial-intelligence. Accessed November 24, 2022.
8 Throughout the empirical chapters of this book we will use quotes from our conversations with the members of the different regional TRANSFORM clusters. The quotes are edited for readability.
9 https://www.openinnovation.regione.lombardia.it/en/ri-in-lombardy/regional-forum/2018-edition. Accessed February 24, 2023.
10 https://www.citizensassembly.ie/en/previous-assemblies/2013-2014-convention-on-the-constitution/. Accessed February 24, 2023.
11 https://www.bandi.regione.lombardia.it/procedimenti/new/bandi/bandi/ricerca-innovazione/ricerca-sviluppo-innovazione/smart-mobility-data-driven-RLF12022027023. Accessed November 30, 2022.

References

Alemanno, Alberto. 2022. "Towards a Permanent Citizens' Participatory Mechanism in the EU." European Parliament Committee on Constitutional Affairs (AFCO). https://doi.org/10.2139/ssrn.4232168.

Arnstein, Sherry R. 1969. "A Ladder of Citizen Participation." *Journal of the American Institute of Planners* 35 (4): 216–24.

Callon, Michel. 1986. "Some Elements of a Sociology of Translation: Domestication of the Scallops and the Fishermen of St. Brieuc Bay." In *Power, Action and Belief. A New Sociology of Knowledge?* edited by John Law, 196–233. London/Boston/Henley: Routledge/Kegan Paul.

Callon, Michel, and Bruno Latour. 1981. "Unscrewing the Big Leviathan: How Actors Macro-Structure Reality and How Sociologists Help Them to Do So." In *Advances in Social Theory and Methodology: Toward an Integration of Micro- and Macro-Sociologies*, edited by Karin Knorr-Cetina and Aaron V Cicourel, 277–303. Boston: Routledge and Kegan Paul. https://doi.org/10.4324/9781315763880.

Courant, Dimitri. 2021. "Citizens' Assemblies for Referendums and Constitutional Reforms: Is There an 'Irish Model' for Deliberative Democracy?" *Frontiers in Political Science* 2 (January). https://doi.org/10.3389/fpos.2020.591983.

Dobson, Andrew. 2012. "Listening: The New Democratic Deficit." *Political Studies* 60 (4): 843–59. https://doi.org/10.1111/J.1467-9248.2012.00944.X.

Dryzek, John S., André Bächtiger, Simone Chambers, Joshua Cohen, James N. Druckman, Andrea Felicetti, James S. Fishkin, et al. 2019. "The Crisis of Democracy and the Science of Deliberation." *Science* 363 (6432): 1144–46. https://doi.org/10.1126/SCIENCE.AAW2694.

Felt, Ulrike, and Maximilian Fochler. 2010. "Machineries for Making Publics: Inscribing and De-Scribing Publics in Public Engagement." *Minerva* 48 (3): 219–38. https://doi.org/10.1007/s11024-010-9155-x.

Goodin, Robert E., and John S. Dryzek. 2006. "Deliberative Impacts: The Macro-Political Uptake of Mini-Publics." *Politics and Society* 34 (2): 219–44. https://doi.org/10.1177/0032329206288152.

Goven, Joanna. 2006. "Dialogue, Governance, and Biotechnology: Acknowledging the Context of the Conversation." *The Integrated Assessment Journal* 6 (2): 99–116.

Grönlund, Kimmo, André Bächtiger, and Maija Setälä. 2014. *Involving Citizens in the Democratic Process*. Colcherster: ECPR Press.

Horst, Maja, and Alan Irwin. 2010. "Nations at Ease with Radical Knowledge: On Consensus, Consensusing and False Consensusness." *Social Studies of Science* 40 (1): 105–26. https://doi.org/10.1177/0306312709341500.

Irwin, Alan. 2006. "The Politics of Talk: Coming to Terms with the 'New' Scientific Governance." *Social Studies of Science* 36 (2): 299–320. https://doi.org/10.1177/0306312706053350.

Jančić, Davor. 2023. *The Changing Role of Citizens in EU Democratic Governance*, edited by Davor Jančić, Bloomsbury Publishing. https://doi.org/10.5040/9781509950850.

Latour, Bruno. 1987. *Science in Action. How to Follow Scientists and Engineers Through Society*. Cambridge: Harvard University Press.

Lezaun, Javier, and Linda Soneryd. 2007. "Consulting Citizens: Technologies of Elicitation and the Mobility of Publics." *Public Understanding of Science* 16 (3): 279–97. https://doi.org/10.1177/0963662507079371.

Mansbridge, Jane. 2010. "Deliberative Polling as the Gold Standard." *The Good Society* 19 (1): 55–62. https://doi.org/10.1353/gso.0.0085.

Michael, Mike. 2012. "'What Are We Busy Doing?' Engaging the Idiot." *Science, Technology & Human Values* 37 (5): 528–54.

Niemeyer, Simon. 2011. "The Emancipatory Effect of Deliberation: Empirical Lessons from Mini-Publics." *Politics and Society* 39 (1): 103–40. https://doi.org/10.1177/0032329210395000.

OECD. 2020. *Innovative Citizen Participation and New Democratic Institutions. Catching the Deliberative Wave*. Paris: OECD Publishing. https://doi.org/10.1787/339306da-en.

———. 2021. *Evaluation Guidelines for Representative Deliberative Processes*. Paris: OECD Publishing. https://doi.org/10.1787/10CCBFCB-EN.

Soneryd, Linda, and Nina Amelung. 2016. "Translating Participation: Scenario Workshops and Citizens' Juries across Situations and Contexts." In *Knowing Governance. The Epistemic Construction of Political Order*, edited by Jan-Peter Voß and Richard Freeman, 155–74. Palgrave Macmillan UK. https://doi.org/10.1057/9781137514509_7.

Stilgoe, Jack, Richard Owen, and Phil Macnaghten. 2013. "Developing a Framework for Responsible Innovation." *Research Policy* 42: 1568–80. https://doi.org/10.1016/j. respol.2013.05.008.

Strand, Roger, and Jack Spaapen. 2021. "Locomotive Breath? Post Festum Reflections on the EC Expert Group on Policy Indicators for Responsible Research and Innovation." In: *Assessment of Responsible Innovation. Methods and Practices*, edited by Emad Yaghmaei and Ibo van de Poel, 42–59. Routledge. https://doi.org/ 10.4324/9780429298998.

Suiter, Jane, and Min Reuchamps. 2016. "A Constitutional Turn for Deliberative Democracy in Europe?" In *Constitutional Deliberative Democracy in Europe*, 1–14. https://www. perlego.com/book/573536/constitutional-deliberative-democracy-in-europe-pdf.

Turnhout, Esther, Tamara Metze, Carina Wyborn, Nicole Klenk, and Elena Louder. 2020. "The Politics of Co-Production: Participation, Power, and Transformation." *Current Opinion in Environmental Sustainability*. https://doi.org/10.1016/j.cosust.2019.11.009.

von Schomberg, René. 2011. "Towards Responsible Research and Innovation in the Information and Communication Technologies and Security Technologies Fields." Publications Office. https://data.europa.eu/doi/10.2777/58723.

Voß, Jan Peter, and Nina Amelung. 2016. "Innovating Public Participation Methods: Technoscientization and Reflexive Engagement." *Social Studies of Science* 46 (5): 749–72. https://doi.org/10.1177/0306312716641350.

Völker, Thomas, and Ângela Guimarães Pereira. 2023. "'What Was that Word? It's Part of Ensuring Its Future Existence' Exploring Engagement Collectives at the European Commission's Joint Research Centre." *Science Technology and Human Values* 48 (2), 428–53. https://doi.org/10.1177/01622439211046049.

Wynne, Brian. 2007. "Public Participation in Science and Technology: Performing and Obscuring a Political–Conceptual Category Mistake." *East Asian Science, Technology and Society: An International Journal* 1 (1): 99–110. https://doi.org/10.1215/ s12280-007-9004-7.

———. 2011. "Lab Work Goes Social, and Vice Versa: Strategising Public Engagement Processes." *Science and Engineering Ethics* 17 (4): 791–800. https://doi.org/10.1007/ s11948-011-9316-9.

5 Citizen science as innovation governance in Catalonia

Stories from the basement

In December 2021, two of the authors of this book (Völker and Strand) rented a meeting room in a hotel downtown Barcelona. The hotel could be said to be stylish in the way that Barcelona is known for: Elegant design and mindful architectural details, well-placed pieces of art within an otherwise mundane, essentially functionalist concrete building. Barcelona herself was less arty than usual, following almost two years of COVID-19 with full and partial lockdowns. Restaurants and cafés were mostly open and could be visited by anyone able to show a valid corona QR passcode. However, public and private offices, universities, and other enterprises – whose workers could work in their homes with their laptops – were still mostly closed. And the normally vibrant streets of Barcelona felt empty.

So what kind of fieldwork can be done when the field is closed? Our solution was to rent a room in our hotel and invite the informants there, for research interviews, interviews that we *could* have performed on Teams or Zoom, but that we were glad we did in person. The way the interviews turned out, they were a reminder after the lockdowns of how profound conversations can become when they take place in a real room with real people and not between pixeled avatars and ghosts. These interviews, together with written materials, recordings, and a host of digital conversations in the years 2020–2022, form the backbone of this story about the TRANSFORM project as it developed in Catalonia; about the idea of citizen science (CS) as innovation governance; about translations of responsible research and innovation (RRI) at the regional level; and most profoundly perhaps, about how to think and enact a desire to transform one's own society.

In contrast to the rest of the hotel, the meeting room was anything but stylish. A windowless room in the basement with cheap furniture heaped together, it felt most of all like an interrogation cell. Under the influence of this ambience, we interviewed and questioned the key participants in the Catalan cluster of the TRANSFORM project about their activities – what they had done, what happened, and how they assessed it – but the conversations roomed much more;

DOI: 10.4324/9781003371229-5

indeed, to anticipate the conclusion of this story, the essence of all the activity was that there *was more,* there was more than what could be seen at first sight. The significance of the activities was wider than their simple description. For instance, from its description one of the project pilots could seem quite mundane – an attempt to involve some citizens of the small Catalan municipality of Mollet del Vallès in an initiative for improved management of domestic waste – but it was *more than that.* It was also an activity with a transformational ambition, an attempt at changing the very culture and practice of governance in this seemingly insignificant suburbia at the outskirts of the Barcelona metropolitan area, making it a showcase for a different possible future in the whole region. To comprehend and appreciate the full value of this little suburban project and its sister pilot at Hospital Sant Pau, one needs to listen to, interpret, and dwell upon the stories of the key participants. We hope that we have been able to render these stories faithfully in this chapter, though shaped by our own interpretations and analyses. However, before getting there, into the thickest stories and interpretations, we will begin with the simple descriptions: The basic idea of what the TRANSFORM project was supposed to be in Catalonia, and why.

"Citizen science in RIS3": The Catalan cluster of TRANSFORM

In previous chapters we have described the overall structure and content of the TRANSFORM project. TRANSFORM was conceived as a project bid. It responded to a particular call for proposals in the Horizon 2020 framework programme of the European Union (EU), namely the SwafS-14 call that was devoted to supporting the "development of territorial RRI," RRI being an acronym for the European policy concept of RRI. The TRANSFORM proposal gathered actors in three European regions – Lombardy, Brussels-Capital Region, and Catalonia – all of them admittedly among the better-developed regions in Europe with respect to research and innovation (R&I). The core idea of the proposal was that three regional clusters would all contribute to the strengthening of RRI (whatever that entailed) in each their region and specifically with regard to the further development of the regional S3 strategies, that is, the smart specialisation strategies. Each regional cluster would focus on a specific methodological approach. In the Catalan case, the regional S3 strategy was the so-called RIS3CAT strategy, and the chosen approach was that of CS. The imagined activities of the Catalan cluster were described in a specific work package, or a "WP" in project jargon, of the TRANSFORM proposal. It was called "WP4: Citizen Science in RIS3 – Catalonia Region."

The high-level objectives of WP4 included (TRANSFORM proposal WP4):

- Embed participatory strategies and CS methodologies in the new RIS3CAT strategy of the Catalonia region, advancing the current efforts to incorporate RRI through the SeeRRI project,

- Transform ongoing RIS3CAT-funded projects from the triple helix towards the quadruple helix, to incorporate bottom-up strategies, co-design customised services, and boost innovation, and
- Set up and carry out a multi-stakeholder engagement to embed public participation and CS in ongoing RIS3CAT projects.

Given the call text from the SwafS-14 call of Horizon 2020, these objectives are not surprising. Indeed, something along these lines had to be offered in order to meet the expectations stated by the call. As a matter of fact, the TRANS-FORM proposal received an extraordinarily positive evaluation (14.5 points, where 15 is the upper theoretical limit) and was selected for EU funding. Yet, a reflection on the level of ambition stated in these objectives compared to the size of the funded project makes it clear how daunting the task was, at least if taken literally. Catalonia is an autonomous region of Spain with close to 8 million inhabitants, with a relatively highly developed R&I sector and 12 recognised universities. Its smart specialisation strategy for the period 2015–2020, the so-called RIS3CAT, counted by its own with a total activity budget of 750 million euros (a considerable part of which was to be funded via the European Regional Development Fund). The Catalan cluster of TRANSFORM and the work foreseen by WP4 of the TRANSFORM proposal and later grant agreement had a total economic value of approximately 250,000 euros and a total of 21 person-months of work, or around 7 months of work in each of the three project years. It was not even a David for transforming Goliat. It was more like a mosquito trying to transform an elephant.

Going beyond the mere numbers and looking into the content of the Catalan smart specialisation strategy, the TRANSFORM objectives did not appear less ambitious. The RIS3CAT 2014–2020 strategy[1] was already a modern one, indeed an elephant and not a dinosaur. The original RIS3CAT strategy document repeatedly stated how the grand challenges of our time necessitate fundamental change in the direction of becoming smarter, more sustainable, and more inclusive. It emphasised the quadruple helix as a key principle. And, as alluded to by the WP4 objectives, there was already an ongoing SwafS-14 project with more or less the same objective for Catalunya, funded in the previous year, namely the so-called SeeRRI project.

Furthermore, below and surrounding the European policy jargon of smart specialisation and quadruple helices, there was the Catalan social and political context with long lines going back to the Spanish Civil War (and before), through decades of oppression and resistance during the Franco regime, towards the Catalan project of building something close to a nation-state after Franco died. A landmark moment of this modern Catalan project was when the President-in-Exile of the Generalitat de Catalunya, Josep Tarradellas, returned to Barcelona and gave his first speech from the balcony of the Generalitat Palace, 23 October 1977. "Ja sóc aquí!" – Now I'm here! – Tarradellas exclaimed and thereby ended the 38-year exile of the Generalitat as the Catalan institution of self-governance, an institution with a history going back to year of 1359.

In 2017, two years before TRANSFORM, the Catalan authorities defied the Spanish state and organised a referendum on the highly contested issue of Catalan sovereignty. Forty years and 4 days after Tarradellas' speech, on 27 October 2017, the Catalan parliament declared the independence of a new Republic of Catalonia. The declaration was annulled and severely punished by the Spanish State, with long prison sentences for its architects, for rebellion and sedition.

While we never – neither in our hotel basement nor in other TRANSFORM settings before or after – heard mention of a connection between S3, RIS3CAT or CS with the political quest for Catalan independence, the social and political context and sentiment was still a fact. Catalonia aspired and aspires to be a modern society with progressive policies. What counts as being progressive may vary but there is always an implicit, sometimes unspoken contrast to the recent past of Spain and the real or imagined foil of Spanish backwardness. The progressive character of the RIS3CAT can be understood in this light: In the Catalan context, R&I policies could be framed as modern, progressive, participatory, and transformational ambitions and attuned to contemporary societal challenges – not merely complying with EU jargon on the matter but going beyond it. What could there be left for TRANSFORM to achieve? Or perhaps the realities of the Catalan society and its R&I ecosystem in particular were more complex and demanding than what could be achieved by the mere discourse of its policies?

Later in the chapter, we shall return to this question about the complexity of the ecosystem. Our answer is going to be affirmative. Meanwhile, at the level of discourse, we may note how TRANSFORM carved out a role for itself by specifying one particular approach, namely that of CS, as its methodological foundation. TRANSFORM Deliverable 2.4 offered a succinct description of the activities of the Catalan cluster of this project:

> The aim of the Catalonia Region in TRANSFORM is to incorporate citizen science as a means of integrating RRI into Catalonia's RIS3CAT 2021-2027, its instruments and the actors of the Catalan R&I ecosystem. TRANSFORM offers an experimentation space that allows the Catalan Government to explore how citizen science could be integrated in RIS3CAT.

> For this purpose, Catalonia Region is developing two citizen science pilot projects in the fields of waste and health. In addition to the pilot projects, the Catalan cluster is developing participatory webinars with the members of the Think Tank with the aim of increasing knowledge about RRI and citizen science and boost the generation of future new projects based on a collaborative framework between stakeholders.

CS was less of a strategic choice than one of identity: The partner in the lead of the Catalan cluster and one of the architects behind TRANSFORM was Rosa Arias, founder and CEO of the small enterprise *Science for Change,* whose primary business model was exactly that of initiating and executing CS projects. With them as partners in the Catalan cluster, they had the Open Systems Research Group at the University of Barcelona, which also focused on CS and participatory research, as well as the Generalitat de Catalunya itself, represented by a public administrator who played a major role in developing RIS3CAT. Many organisations and individuals ended up playing a role in the Catalan cluster of TRANSFORM and its activities. The main strategic actors, however, remained the same throughout the project: Science for Change, the Open Systems Group, and the Generalitat.

Why citizen science?

Before we continue on our journey through the Catalan cluster, it may be useful to briefly recapitulate what CS is and may be. The term itself is relatively new. It is found in academic literature since the early 1980s, but its use was rare until the 2000s (Kullenberg and Kasperowski 2016). In 2014, it came into prominence by being included in the Oxford English Dictionary (OED).[2] The OED definition reads:

> **citizen science** n. scientific work undertaken by members of the general public, often in collaboration with or under the direction of professional scientists and scientific institutions.

Exactly what counts as CS, what it is and what it is good for, are issues where opinions differ and ideas are contested. Part of the disagreement can be seen as ideological and it resembles rather closely part of the disagreements about what RRI is and should be, revolving around the questions about the legitimate role of civil society in the governance and practice of science. Should society "speak back to science," as suggested by Gibbons et al. (1994)? Should science become more democratic?

These disagreements have attracted attention from scholars within the field of science and technology studies (STS); indeed, their original formulation and analysis were due to STS scholars such as Alan Irwin (1995, 2001). If we follow Kullenberg and Kasperowski (2016), however, and look at research papers in which the term "citizen science" is used, the scholarly discussions about the term itself amount to a minor portion of the corpus. The majority of papers that speak of CS are contributions to natural science, in particular within ecology, environmental sciences, biology, and geography, and to some smaller extent fields such as computer science and astronomy. The studies engage with the term CS as a label to describe their research approach. Often, the approach is one of natural science with contributions from "citizens," which may be observations, use of sensors, data analysis, lending computing power from personal computers, etc. It is fair to say that the contributions tend to have the character of research assistance. Moreover, as is easily observed from the OED definition, the concept of CS rests on some kind of demarcation between science/scientists and non-science/(lay) citizens which – for the concept to make sense – is not the philosophical demarcation between justified and less justified claims to knowledge but rather an institutional one. Citizens may contribute to science even if they are not *professional* scientists with proper scientific jobs (and salaries). "Even if" signals that their status and the terms of the collaboration with the professional scientists are not equal, though. To apply a proverbial Orwellian formula, all scientists are equal but some are more equal than others, namely the professional ones.

In parallel with and to some extent preceding the exponential growth in scientific research involving the practices of CS just described, there is the scholarly literature around it, arguably with STS and Alan Irwin's scholarship in the centre. In this literature, CS practices are critiqued, for instance, for being exploitative or co-opting, but also promoted as an opportunity for a type of public engagement with science that holds a potential for citizen empowerment and the democratisation of science. In this regard, CS belongs to a family that includes several fields of practice. An older and possibly more prominent family member is the tradition(s) of participatory action research that has roots back to Kurt Lewin in the 1940s and that proliferated in several continents in the 1970s also as part of civil rights movements. Other neighbouring concepts are those of community-based research, public participation, science cafés, participatory technology assessments, social/participatory innovation, living labs, etc. The growth of the institutional practices of research ethics is perhaps also related to the same family of concepts. Indeed, one way of looking at the invention and incorporation of the concept of Responsible Research Innovation is to see it as an attempt to gather many of these quite diverse strands of ideas and practices into an over-arching concept that could coordinate efforts, communities, and policies that all ultimately seek a better alignment between scientific practices and institutions with civil society and its ethical and political values. In the European Commission definition of RRI, CS would then naturally fit within the so-called policy key of public engagement, and also with the later policies on Open Science.

A central issue at stake, then, for CS and RRI and not the least their intersection, is the degree to which the activities or practices concerned actually hold an emancipatory or empowering potential for society to speak back to science and for citizens to shape the scientific practices and projects or technological innovation processes to which they contribute. In the absence of such a potential, CS may become just another instance of what Arnstein (1969) criticised as tokenism or de facto non-participation, simply serving purposes of informing, placating, manipulating, and co-opting the public. This risk is present for CS but also for any other contemporary practice within the family of public participation and/or engagement with science and technology (Völker and Pereira 2023).

Returning to the Catalan cluster of TRANSFORM, two of the main actors – Science for Change and the Open Systems Group – were practitioners and indeed experts within the field of CS. Both of them were associated with the Barcelona Citizen Science Office, which was created by the city council of Barcelona in 2012. This office as well as our two protagonists all recognise the diversity of citizen projects. On the website of the Citizen Science Office, four levels of CS are described: (1) Crowdsourcing, (2) distributed intelligence, (3) participatory research, and (4) collaborative research.[3] The higher the level, the more deeply engaged are the citizen participants and the higher their responsibility. The website, our informants, and indeed the scholarly literature on the field agreed, however, that the maximum level is not necessarily always optimal. It depends on the nature and purpose of the project. Still, the emancipatory and empowering purpose of CS was key to the actors in the Catalan cluster, and this idea was very vocal within the Open Research Group:

> … they just want people engaged there and just answering questions and get the aggregated data and that's it. That's not what we wanted at all. And that's one of the tensions that is fully present in citizen science, I would say.
>
> (Int_04)

In terms of the levels just mentioned, their ambition was that of participatory research where citizens are invited as epistemic actors who might possibly be the real experts on some parts of reality that the professional scientists did not even know about. However, the ambition went beyond that of improving science by enriching it with more epistemic actors. It went beyond the demarcation of science and society as such and, as the name indicated, viewing society as *open systems:*

> OK, so we can do a very nice project with a lot of data and so on but what is next? What is going to happen with that?
>
> (Int_04)

What is going to happen in the community? How can good, progressive processes be initiated and catalysed by for instance CS, as a means for public experimentation that arises out of concrete issues? In the words of the Citizen Science Office of Barcelona, how can CS be a vehicle for "improving reality"?

The vision of improving reality via CS and RRI was also vocal in Science for Change:

> (…) you are actually trying to solve real challenges with people, involving people who are, in the case of citizen science, actively contributing to

the research, the results [...] and the final product of that would align with society for sure because you are working with society but also it will be much more innovative.

(Int_01)

It will be more innovative because the diversity of actors involved will (hopefully) lead to a diversity of ideas and approaches. Such a vision of CS implicitly places it on the higher ambition levels for participation and collaboration, whether placed on Arnstein's famous ladder of participation or within the more mundane four-level taxonomy used by the Barcelona office. While practices may vary, the flavour of citizen science as it was conceived in Catalonia shows its ideological connection to Irwin and the critique of the deficit model and back to the traditions of participatory action research.

It can be argued that RRI conveniently fits this particular context. The status of RRI as a cross-cutting principle of Horizon 2020 provided funding opportunities as well as EU-level political justification of whatever could be recognised as RRI. In Science for Change, RRI and citizen science were considered as mutually conducive of each other. In their conception, citizen science is a practical approach that can be used to implement the normative values of the RRI principle. Conversely, the normative force of RRI gives directionality to citizen science so that it indeed aligns well with society and takes the desired emancipatory and innovative character. In fact, in Science for Change, even the so-called RRI policy keys of the European Commission (gender equality, ethics, public engagement, science education, open access/open science, and possibly governance) were considered to have instrumental value in that they could be applied to citizen science projects as questions or challenges. One may address the gender dimension of the project, the ethical challenges, etc.

In short, several longer threads can be identified in the textile of the Catalan cluster of TRANSFORM and its notion of citizen science: The legacy of participatory action research that in a way was reinvented in parts of citizen science; the many different forms of public engagement with science and technology that were assembled into the umbrella concept of RRI; and the social and political project of creating a modern and progressive Catalonia by "improving reality."

CS has been blooming in several parts of the planet. Although the worldwide CS growth shares important commonalities and seems to have a general trend everywhere, it is also true that some regions and cities have more CS activity than others. This can be attributed to cultural particularities or to some conditions that are more favourable to these participatory research practices. In places like Europe, countries appear to be more active than others if one consults the number EU funded projects. Spain and more particularly Catalonia region show a relevant number of projects with the term "citizen science." Reasons related to these disparities are however difficult to capture with more figures. Catalonia has yet no specific R&I policy (with or without funding) on CS. The term appeared for the first time in a recent and unique Generalitat de Catalunya (regional government) funding call exclusively related to the COVID pandemic (2021), but it has not appeared again in any regular call. The presence of CS term can be explained by the fact that the call particularly welcomed transdisciplinary approaches and actionable knowledge. The growth of a number of researchers from several universities and research centres in CS during the last decade can neither attributed to specific institutional support nor recognition. It is however important to mention that the R&I regional strategy is starting to be attentive to CS. In 2022, the Generalitat scientific and innovation research policies from different departments have opened different spaces and workgroups to structure new strategies, to deliver reports and policy briefs where CS is present or is about to be present. Policymakers have grounded their discourse around CS as one possible way to untap innovation, increase social inclusion in sustainability transitions or augment democratic values within the broad Open Science framework. A similar phenomenon is also happening within most of the public universities. Since 2022, they are organising internal networking events or training sessions to their researchers while starting to showcase their own CS projects on their public websites thus seeing CS as one of the ways how the institution is contributing and sensitive to societal problems. In any of these cases, there is still no sign to give official recognition or any reward to those researchers involved in CS practices, but it is already under discussion. However, this slow response has not stopped the growing trend, and this could be possibly attributed to the perception of involved researchers that they need to reconsider their own research agenda by the further involvement of citizens. Additionally, there are other conditions that are invigorating CS locally. Barcelona municipality has been offering networking spaces to enrich the community, showcasing the existing projects in public and massive events, or running specific programmes with schools through the Barcelona Citizen Science Office and guided by the Barcelona Science Plan 2020–2023. Other municipalities and other regional institutions have also seen in CS to reimagine contexts and their functionalities such as public libraries, museums, cultural centres, and community centres. Also, one may say that civil society is also playing a very important role. In Catalonia, there is a strong tradition of solid, but small, organised civil society initiatives. There is a strong cultural background deeply enrooted in the Catalan society. Civil society organisations are indeed currently becoming partners of some CS projects or they have been using CS practices without using the term for decades to monitor the quality of rivers or building a census of homeless people living in the streets of Barcelona. They are seeing CS as a practice aligned to their own goals and as a way to give them a louder voice in societies. (*Josep Perelló, University of Barcelona*)

RIS3CAT: Third-generation innovation policy at the regional level

Still, the mandate of TRANSFORM was not simply to perform or promote CS. The stated purpose of TRANSFORM was to integrate the principle of RRI into RIS3CAT, the smart specialisation strategy of Catalonia, *by use of* CS. It should, to once again quote TRANSFORM Deliverable 2.4, offer "an experimentation space that allows the Catalan Government to explore how CS could be integrated in RIS3CAT."

In this way, CS became the translation of RRI in a double sense. The desired end result, which was to integrate RRI into RIS3CAT, was translated as the integration of CS into RIS3CAT. And the means by which to do so, that is, the RRI intervention of TRANSFORM, was also conceived as CS. The explanation or rather, the theory of change by which this would be possible, was one of exemplary learning. TRANSFORM would *do* CS and thereby *show* the Generalitat that CS could be well integrated into RIS3CAT. There are several subtleties to be commented in this double translation, and they will be illustrated by the empirical detail below. However, we may already now observe that the implicit theory of change presupposed that the CS achievements of TRANSFORM indeed would become exemplary. It is hard to see how they could serve the transformational purpose if they happened to fail or be seen as weak or unpurposeful.

Another subtlety is the issue of how an experience or achievement becomes perceived as exemplary. What kind of work is needed for something to shine and to be seen as a good example, both by those shining and those seeing?

And finally, one could ask about the validity of the translation of RRI into CS. There were at least two answers to this question, answers that were not mutually exclusive. Firstly, within the described "progressive" context, the ideal of CS was connected to public engagement and the idea of empowering citizens to speak back to science and shaping R&I. Secondly, in the Science for Change vision already mentioned, the CS to be applied was infused with the values embodied by the RRI policy keys of the European Commission. TRANSFORM should do "RRI-laden" CS.

From within what actors in Horizon 2020-funded RRI projects sometimes called "the RRI bubble," that is, the loosely organised epistemic network of researchers, consultants, and what in technical EU jargon would be called "beneficiaries" of such project grants, the assumptions made in this story of integrating RRI in RIS3 strategies by use of CS makes perfect sense. In general, outside of the RRI bubble, the argument is not self-evident at all. Within the sphere of smart specialisation strategies, there is a diversity of ideas and practices. While some of them would be compatible with the idea of CS for transformation, for others that idea is rather alien.

In Chapter 3, we presented the stated rationale of smart specialisation and RIS3. It may be recalled that the RIS3 requirement that regions should develop their own smart specialisation strategies was introduced as a part of EU's cohesion policy. The explicit objective was economic transformation to boost economic growth and job creation by focusing – specialising – on regional strengths in industry and R&D that could result in competitive advantage. From the Brussels perspective, this could ensure good use of EU subsidies. Regions could continue to receive money from the EU via the ESIF – the European Structural and Investment Fund, of which the European Regional Development Fund is a part – but only if they presented objectives on and a strategy for building competitive advantage in the form of a RIS3 plan. The policies and more practical documents, such as the RIS3 Guide,[4] explained how such a plan could be set up and organised.

Otto von Bismarck famously compared law-making to the making of sausages. Both types of processes tend to produce composite results. Smart specialisation can be seen as its own bubble with its own complex and heterogenous epistemic network, and RIS3 policy documents are highly composite. On one hand, the macroeconomic perspective is strong; it is a growth policy. On the other hand, the growth is supposed to be not only smart but also sustainable and inclusive. The RIS3 guide also contains advice on social innovation, on the need for stakeholder participation, and on innovation within a quadruple helix mindset. Taken as text, the RIS3 guides and factsheets cannot help leaving the impression of cognitive dissonance, witnessed in for instance the use of words such as "citizen." In the official RIS3 factsheet, citizens are never mentioned. In the 2012 RIS3 Guide, the word is used a dozen times. Sometimes the citizens are envisioned as active participants; at other times they are imagined as users of innovations and consumers of products; and occasionally they are simply objects to be observed and governed, as in the following description of research infrastructures: "Their know-how helps European industry develop new pharmaceuticals and high-performance materials, monitor the earth's oceans and air, and track the changing social attitudes and behaviour of our fellow-citizens" (European Commission 2012, 74). And if we look at the endorsed methodology to monitor how the regions are performing, the so-called Regional Innovation Scoreboard, there is nothing there that has a taste of RRI, CS, or the need for civil society to speak back to science. The indicators all belong to a combined technocratic and macro-economic paradigm, measuring investments, numbers of patents and publications, percentage of population with higher education or digital skills, etc.

The advantage of the dissonance and ambiguity in the documents, from a practical political perspective, is that quite diverse regional approaches can be justified and find endorsement. It was fully possible to write a RIS3 plan that is little more than a business-as-usual second-generation innovation policy strategy that addresses systemic features of the regional innovation ecosystem, perhaps with some attention given to sustainability challenges. However, it was also possible to find justification for more radical approaches, not the least with the growing attention to transdisciplinarity and wider concepts of transformation in OECD policy. In scholarly literature on regional development policies, the idea of seeing S3 as an opportunity for wider transformations has gained some traction (Fitjar, Benneworth, and Asheim 2019; Coenen and Morgan 2020).

The RIS3CAT was such an example of a more radical approach, aligning itself not only with the concepts of quadruple helix and social innovation but also with the diagnostic and rationale of third-generation innovation policies. In addition to the more conventional (and mandatory) S3 instruments, RIS3CAT contained a set of foci for the development of public policy which included digital agendas, entrepreneurship, and education but also eco-innovation and "non-technological innovation." Social innovation was highlighted in the latter as processes to address neglected social needs and to develop new social relations and new forms of collaborations in society:

> The processes involved in social innovation result in learning, commitments and transformations that affect the local sphere, and the construction of processes should be based on the participation of local stakeholders, empowerment and citizen engagement.
>
> (Generalitat de Catalunya 2014, 51)

In sum, one can observe that the objectives of the TRANSFORM project and the orientation of RIS3CAT were both compatible and incompatible. They were compatible because RIS3CAT already was situated within third-generation innovation policy and within the mindset of participation, empowerment, and transformation. In this sense, CS was a plausible addition. At the same time, the expressed objective of TRANSFORM – to transform RIS3CAT by integrating CS (as a form of or proxy for RRI) – was perhaps more relevant for satisfying the SwafS call under which TRANSFORM was funded than for RIS3CAT itself. The underlying normative rationale of RRI could be seen as already present in RIS3CAT. Indeed, the very idea of a small EU-funded action as a change agent for a public policy, even if it were possible, raises principled, democratic issues, at least if the change is construed in terms of conventional intervention logic, asking for the project to have "impact" on public policies. In the Catalan cluster of TRANSFORM, however, the problems of intervention logic were not really an issue because all partners operated within mindsets of network governance – two partners specialising in the more empowering end of CS and the regional authority partner (the Generalitat) acting in a third-generation framework. As summarised by one of the Generalitat representatives: "Projects don't change policies." Projects such as TRANSFORM take place within a complex ecosystem of actors and contribute to the governance that takes place within this complexity. They do not govern the complexity by themselves. Projects neither can nor should change policies, and the success of such projects depends on understanding this in practice while somehow satisfying the expectations of the European Commission who asks for "policy impact" of the project. This is yet another instance of translation, from the imaginary of intervention logic to the reality of network governance (and sometimes back to the imaginary, by the use of indicators and other reporting devices).

The pilots: Citizen science in waste management and health management

From the analytical perspective of network governance, a lot may be going on in and around a project, such as informal negotiations and deliberations, learning processes, and many other things. We shall return to this later. From a more formal project perspective, however, the Catalan cluster of TRANSFORM undertook two types of activities as foreseen by the project proposal and the grant agreement, namely the "think tank" and the local pilots. The project established a Catalan RRI think tank with a variety of members recruited from the regional innovation ecosystem. It also ran meetings with the think tank throughout the project period. Furthermore, the Catalan cluster designed and undertook two pilot projects within the greater TRANSFORM project that sought to explore and demonstrate the value of CS methodologies. One such pilot was already foreseen in the proposal phase of TRANSFORM, namely an initiative that introduced CS elements in an ongoing policy process on waste management in a small Catalan municipality called Mollet del Vallès. The other pilot was defined during the project period and originated from deliberations in the think tank. Its topic was treatment of endometriosis at the Hospital Sant Pau in Barcelona, and it set up a participatory co-creation process with patients and medical staff at the hospital.

Waste collection in Mollet del Vallès

One of the Catalan TRANSFORM pilots applied a CS approach to work with young citizens (secondary school pupils) in the suburban town Mollet del

Vallès and several departments of the municipality with the aim to improve local waste collection practices. Together with the youngsters, the local authorities, and students from a Catalan university, the project group co-designed a serious game, an interactive digital waste game called Dilemma-R. Subsequently, the waste game was used by the secondary school students to elicit preferences for waste collection practices among citizens in the town. The exercise was deemed successful also in the very practical sense that it yielded results that were incorporated into the next public tender for waste collection in Mollet del Vallès.

For a glossy leaflet about the TRANSFORM project, we could wrap up the story about the waste game pilot here. It was perceived as a success, it achieved its concrete objectives, and it provided the Generalitat with a proof of principle that CS methodologies can deliver the expected results. In that sense, it contributed to the overall purpose of exemplary learning towards the integration of CS in the next RIS3 strategy of Catalonia.

Furthermore, in the language of RRI, the pilot could be described as aligning the local technical waste collection system better with the values, needs, and demands of the local population. The role of the young participants was above all to serve as ambassadors and door openers to the wider population of the town. In terms of the RRI AREA framework, the game created a space for reflection, engagement, and deliberation, to which the municipality proved itself as responsive by virtue of policy uptake of the results of the deliberative exercise. In terms of the notorious "policy keys" of the European Commission, there was public engagement and perhaps also some science education.

And yet, someone unfamiliar with the project and its context might wonder what suburban waste collection has to do with regional innovation or RRI for that matter. Was the waste game an innovation, and if so, a notable one? Was this technoscience? Was there even research? The exercise looked quite different from the type of RRI initiative and practices that we will present as a contrast in Chapter 8, in consortia and endeavours that have their core in academic research environments within biotechnology. Also, bearing for instance the indicators of the Regional Innovation Scoreboard in mind, it could be hard to see how the waste game pilot had anything to do with the sort of thing that the scoreboard aims to measure. For instance, there was no apparent business model for the waste game and seemingly no job creation involved (other than the work created by the TRANSFORM grant agreement itself). One could easily imagine innovation economists and policymakers discard the glossy leaflet and dismiss the story about Mollet as neither relevant nor noteworthy.

To appreciate the real significance of the waste game pilot, one needs to return to the regional context, the transformative ambitions of RIS3CAT, and the work modes of the Generalitat in their implementation of RIS3CAT.

We will begin by setting the scene in Mollet del Vallès. Mollet is a town with 51,000 inhabitants (2021) located less than 20 kilometres East of Barcelona, in the county of Eastern Vallès. Its history goes back to the 10th century. However, the character of the town has changed with the growth of Barcelona and its metropolitan area. Mollet is not part of the formally defined Metropolitan Area of Barcelona but belongs to the so-called second "corona" or (half-)crown around it. Still, to travel by train from Barcelona to Mollet is a continuous experience of urban and suburban spaces and infrastructures. As many of the other municipalities surrounding the big city, its challenges are shaped by these infrastructures, with the AP-7 highway at its North border, railways criss-crossing the space, an enormous car park for unsold cars next to the town, and a population that doubled between 1960 and 1970, and then again between 1970 and 1990. In the latter years, however, the population size has been more or less stable. The area is built up and urbanised to such an extent that it is difficult to see how it can expand further, unless its agricultural and industrial areas to the North and East are also urbanised.

For the TRANSFORM purpose, there were several features of Mollet that could be seen as attractive. First and most important of all, the local authorities were interested in participating in the pilot. This might have been a contingency. On the other hand, this could also be due to a more general motivation related to the social and geographical context itself. Mollet is one example of these towns around Barcelona that now find themselves within a huge urban landscape and with relatively newly shaped populations and that share both the desire and need to further develop their identity as communities and "improve reality," to paraphrase the Barcelona Citizen Science Office. Perhaps less than a top-down strategic consideration, the choice of going to Mollet emerged out of the interactions of the Catalan RRI think tank, to be described below.

Mollet turned out to be a happy choice. In the city of Barcelona itself, perhaps, there would be a longer history for CS and for public participation exercises, but on the other hand, Barcelona in its entirety would be too big for a small project such as TRANSFORM to manage. In Mollet, the project partners could get to interact with the relevant actors within the technical and financial services of the municipality. Moreover, they were able to cover the town by reaching out to its secondary schools. Indeed, it was highlighted by one of the partners that participation mechanisms that go through the school system make for geographical coverage, simply because there are schools where people live. By making the youngsters the "ambassadors" of the project, gates were opened to the entire community through their parents and other relatives and neighbours.

At the same time it could even be argued that the short history of Mollet as a society and community in its present form, with a large influx of inhabitants little more than one generation ago, indeed would provide a stronger proof of concept of CS methodologies than in a community with an older civic texture.

The topic of waste was already highlighted in the TRANSFORM project proposal as a well-suited one, even before it had been decided where the pilot would be located. Still, for CS to make sense in the TRANSFORM context, it would have to fulfil the objective of engaging with actual problems perceived by the citizens. As one interviewee told us, "it would be nice if we can work with somebody's real challenges" (Int_01). This in turn also means that the particular situation – the regional context – in which this pilot takes place strongly shaped what was being done:

> So, what we want to do with the game is to co-create with people like the ideal waste collection system for their neighbourhood. And then you of course need to be able to implement that. So also people in charge needs to be flexible and understand that maybe not all one solution fits all. That maybe you need to have different solutions for the different neighbourhoods.
>
> (Int_01)

In the sense that engagement was achieved by means of the waste game, it may be concluded that the issue of waste collection found interest in this community. Indeed, in the streets of Mollet, one can observe signs of the local authorities giving attention to waste separation, see Figure 5.1.

We were surprised to see mini-recycling stations such as the one depicted, with separate entries for cork, inkjet cartridges, LED lamps, batteries, small electronics, and others. The outcome of the pilot was not "smart" in the sense of making these solutions more digital or more dependent on solutions delivered by technoscience. Rather, what the pilot tried and what it did was to improve transversal communication and collaboration in the municipality of Mollet del Vallès. Young and old citizens got to communicate with the municipality in new ways on a public issue of interest. And perhaps equally important, the pilot developed transversal communication between the technical, financial, and educational services. This may sound commonplace, but it was not perceived as such in the local context and it required a long-term process of developing mutual understanding and a shared language.

In terms of RRI, we can see this as a translation of RRI that aims for responsiveness at the community level of knowledge production. In terms of RIS3CAT, it was a proof of principle that a CS approach could be conducive of non-technical, social innovation aimed at developing and deepening social relations within the community on an issue of public interest, for the common good, but not without potential controversy. This may not be perceived as important if the goal is short-term boosting of the Regional Innovation Scoreboard. However, it is more important if the goal is to address transformation failures in regional development, in light of third-generation innovation policy.

Figure 5.1 Photo from Mollet del Vallès, picture taken by Roger Strand, 2022.

Endometriosis in the first person

Based on the interactions and ideas that emerged in the Catalan think tank (see below), an additional pilot was defined with the objective of improving services for the diagnosis, care, and support for women suffering from endometriosis. Endometriosis is a disease connected to the female reproductive system. It is associated with immense pain and, while being a serious condition, it is often diagnosed several years after the onset of symptoms.

The pilot was called "Endometriosis in the First Person" and employed a CS approach with patients and medical staff at Hospital Sant Pau in Barcelona. The Catalan Agency for Health Quality and Assessment was also part of the initiative. The goals of the pilot activity were twofold: Co-creation of recommendations to inform a new protocol for endometriosis care in Catalonia, and capacity-building and improved transversal communication and collaboration between public administration, health personnel, and patients.

As with the waste pilot in Mollet, the participants of the endometriosis pilot deemed it a success. Recommendations were indeed agreed upon and presented in a policy brief (Salas Seoane, Botsho, and Iannitelli 2022). The brief presented key dimensions of the qualitative findings from the participatory research conducted. They contained rich accounts of how it is to live with endometriosis, in terms of psychological, social, and physiological experiences and their strategies to cope with the illness. Among the key dimensions were the experience of gender stereotypes and negligence in medical care when trying to get help. The recommendations focused on awareness raising, early diagnosis, and improvements in health care by better involvement of the patient in the clinical decision-making (see Figure 5.2). As far as we could observe, all actors expressed enthusiasm about the pilot's ability to energise the communication between and among the patients and the health care personnel:

> [I]n her team there were some other people working on that and she saw a very good opportunity to advance in a different way on this. To talk with the patients and involve them in all of the process. And they are super happy and in fact they are changing the protocols in the hospital.
>
> (Int_01)

The activity thus combined a conventional research interest with a willingness to do things differently by involving the patients. What we see here is the application of CS to gather information on the needs of a vulnerable group of people and then involve members of this group in the creation of an improved health service protocol. At the time of writing (2022), there were prospects of funding and thereby continuing the process further also after the end of TRANSFORM as an EU-funded project. Just as with the waste pilot, it could be concluded that the endometriosis pilot both reached its local objectives and

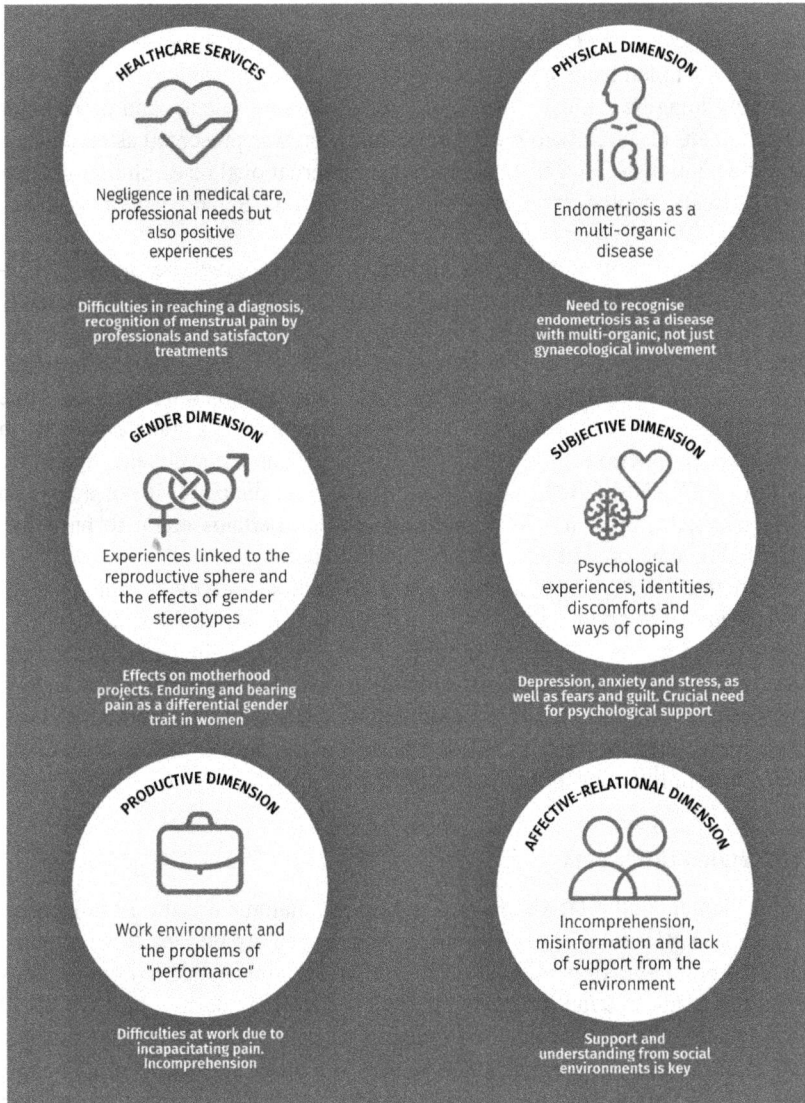

Figure 5.2 Facsimile from the endometriosis policy brief (Salas Seoane, Botsho, and Iannitelli 2022, 5)

presented a showcase for how CS methodologies could be applied, in this case in a research process coupled with public innovation in the health care sector. RRI dimensions of reflexivity, inclusion, and responsiveness could be invoked in the description, as well as policy keys of public engagement and not the least gender equity.

And yet, for potential actors in the S3 bubble who are unfamiliar or uninterested in the value-based discourse of RRI, the pilot could appear as rather remote or uncoupled from smart specialisation as a growth strategy. Additionally, one could imagine critical voices from within nursing science and other fields of health care research who might argue that what was presented as results and recommendations is all well-known from the international research literature on endometriosis and that one does not necessarily need CS to do qualitative health research.

The latter point may not be so relevant at the local level, however. One informant from the hospital described to us that the TRANSFORM pilot actually made a significant contribution simply by creating an arena for dialogue between patients and doctors. Indeed, the informant described a context in which there was little communication even among gynaecologists treating the same patient, especially if they worked at different hospitals. To involve patients in knowledge production signified a huge step in such an environment. Again, we see how a TRANSFORM pilot initiates transversal dialogue, if not across an R&I ecosystem, then in a local environment that perhaps ought to have had more dialogue before but as a matter of fact did not.

From the RIS3CAT perspective, the pilot demonstrated concrete steps towards results that promised an "improved reality," in this case for women with a painful and much-neglected condition. Furthermore, it showed that concrete initiatives could emerge out of a construction such as the Catalan think tank. It provided proof of principle that the think tank could generate concrete change, as it demonstrated the value of a structure and not just a method. Let us therefore move on from the pilots to the think tank itself.

The Catalan think tank

A main element of the TRANSFORM project in Catalonia was the establishment of a Catalan RRI think tank with more than 50 representatives from regional and local policy-making organisations, companies, and academic organisations belonging to the Catalan R&I ecosystem. It was formed during the COVID-19 lockdown and had to base itself on online platforms from which it conducted a quite extensive set of meetings and webinars, organised by Science for Change. Originally, the think tank was envisioned with fewer participants. Due to high interest, however, it was decided that the think tank should increase its size.

The initial focus of the think tank was to build capacity and interest in RRI among the participants. Next, it proceeded gradually to map and develop common interests for RRI activities. Various possibilities for joint initiatives were explored, and the most concrete outcome was the endometriosis pilot at Hospital Sant Pau. Furthermore, although the waste pilot was anticipated already in the TRANSFORM project proposal, the localisation into Mollet del Vallès happened through the interactions in the think tank.

As it happened, the think tank came to serve several purposes. At a straightforward level, it likely led to increased awareness and knowledge of RRI and participatory methodologies among its members, as was corroborated by documentation from the webinars and meetings. Less certain but not unlikely, it may have had a further impact on the level of RRI awareness and knowledge in the wider ecosystem – that is, if the think tank participants informed their colleagues and other actors about their experiences. Next, the endometriosis pilot was – as mentioned – a proof in principle that the think tank could serve as a generator of such initiatives. This was particularly interesting for Science for Change, whose business model and raison d'être indeed is to perform CS projects. Importantly though, the think tank was also an initiative to connect representatives of institutional actors across the R&I ecosystem with an actual or potential interest in RRI or more generally the transformative agenda of RIS3CAT. From the point of view of the Generalitat, or more specifically its General Directorate of Economic Promotion and Regulation, the think tank was an experiment to improve connectivity and new patterns of collaboration within the quadruple helix, and it served to extend the relevant ecological network and fortify its transformative character. To do so, it was important to emphasise the think tank as an arena for actors with that interest, an emphasis that was more general and extended beyond the think tank:

> People who want to transform, I help them. If they don't want to transform, they don't care and I don't care. […] I tell everybody if you want to transform, you are in the right room. If you want money to do the same, next door please.
>
> (Int_03)

From this perspective, it was also natural to expect that some actors disappeared as the work in the think tank developed. Not everybody could be expected to have a real interest in RRI or transformative change, or even if they were interested, to see how to engage with RRI in their own work. The aim was not to keep the think tank as big as possible but rather to crystallise and stay with a network of dedicated individuals who shared the overall ambition.

In terms of the TRANSFORM proposal and grant agreement, the think tank was at the same time both less and more ambitious than the original objectives. It did not "transform ongoing RIS3CAT-funded projects from the triple helix towards the quadruple helix" or "embed public participation and CS in ongoing RIS3CAT projects." People were not invited nor took part in the think tank as objects to be transformed. Such grant agreement objectives that are formulated in an intervention logic remain quite disconnected from the actual context. Instead, it created and nurtured a new network of actors inside the R&I ecosystem. These actors got to learn about RRI in the smart specialisation context, and importantly they got the opportunity to interact with each other and people who shared their interests and value orientations. It was set up to stimulate agency.

From the point of view of the Generalitat, then, the continued existence of the think tank after TRANSFORM was less of a priority. It had served its purpose. We may contrast this point of view with the interests of a service provider such as Science for Change, for which the think tank could be an instrument to continue to tap into the needs and interests of the local network.

Doing RRI as citizen science and "more than that" in Catalonia

What did the Catalan cluster of TRANSFORM achieve? Did the project fulfil its objectives? We may recall the TRANSFORM WP4 high-level objectives:

- Embed participatory strategies and CS methodologies in the new RIS3CAT strategy of the Catalonia region, advancing the current efforts to incorporate RRI through the SeeRRI project,
- Transform ongoing RIS3CAT-funded projects from the triple helix towards the quadruple helix, to incorporate bottom-up strategies, co-design customised services and boost innovation, and
- Set up and carry out a multi-stakeholder engagement to embed public participation and CS in ongoing RIS3CAT projects.

Above we presented the pilots and the think tank, which in a strict sense cannot be said to have "transformed ongoing RIS3CAT-funded projects" but in other, alternative ways to introduce RRI and CS into Catalan society. As for the first and boldest objective, in November 2022, one month before the end of the TRANS-FORM project, the Catalan Generalitat approved their new RIS3CAT strategy, the so-called RIS3CAT2030 (Generalitat de Catalunya 2022; Figure 5.3).

The table given in Figure 5.3 summarises the envisioned workings of the new strategy: "RIS3CAT will promote transformative and RRI with impact on the quality of life of the persons in the territory." It will do so by drawing on generic technologies and new digital and technological industries but also by shared agenda-setting – all of this is quite similar to the previous strategy. The high-level reference to RRI is new, however, and the shared agendas include "a system of education and knowledge production which is reflexive, anticipatory, inclusive and responsive" – a direct reference to the AREA framework of responsible innovation. Further down in the text, CS is explicitly mentioned as an instrument of change:

> More concretely, the development of citizen science will be promoted, of science at the service of the persons and science done together with the persons. To do so, the social and cultural actors will be involved in the identification of the challenges of the territory and the proposal and design of research and innovation projects that may provide solutions to the problems that concern the persons.
>
> (RIS3CAT2030, 14)

Quadre 1. Esquema general de la RIS3CAT 2030

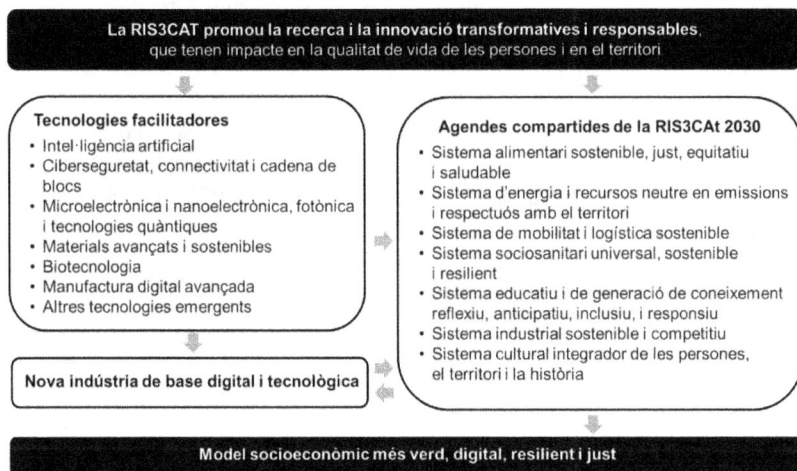

La RIS3CAT promou la recerca i la innovació transformatives i responsables, que tenen impacte en la qualitat de vida de les persones i en el territori

Tecnologies facilitadores
- Intel·ligència artificial
- Ciberseguretat, connectivitat i cadena de blocs
- Microelectrònica i nanoelectrònica, fotònica i tecnologies quàntiques
- Materials avançats i sostenibles
- Biotecnologia
- Manufactura digital avançada
- Altres tecnologies emergents

Agendes compartides de la RIS3CAt 2030
- Sistema alimentari sostenible, just, equitatiu i saludable
- Sistema d'energia i recursos neutre en emissions i respectuós amb el territori
- Sistema de mobilitat i logística sostenible
- Sistema sociosanitari universal, sostenible i resilient
- Sistema educatiu i de generació de coneixement reflexiu, anticipatiu, inclusiu, i responsiu
- Sistema industrial sostenible i competitiu
- Sistema cultural integrador de les persones, el territori i la història

Nova indústria de base digital i tecnològica

Model socioeconòmic més verd, digital, resilient i just

Figure 5.3 Table from the new regional smart specialisation strategy of Catalonia (Generalitat de Catalunya 2022, 12).

Viewed at a distance, then, the mission was accomplished: RRI and CS were included in the regional smart specialisation strategy.

And yet, just as with the individual pilots, there is more to this story. When we examine the case closely the success that we saw at a distance becomes blurred if not superficial, while other subtler achievements come into focus. To begin with the question of "mission accomplished," one could argue that even if the verbiage of RRI, the AREA framework, and CS was not prominent in the previous RIS3CAT strategy, the underlying ideas were already present, as described earlier in this chapter. The specific mention of the methodology of CS could still be important, though, not the least for the Catalan actors who are specialists within CS.

More to the point, perhaps, one could ask if the inclusion of RRI and CS rightly could be attributed to TRANSFORM, or at least the activities of TRANS-FORM. Did the TRANSFORM project "impact" the Catalan Generalitat so that it changed its smart specialisation strategy? To stay with the impact metaphor, were the TRANSFORM activities a kind of projectile that hit the Generalitat with sufficient momentum to shift its course?

Without depreciating the achievements of TRANSFORM, we believe the projectile and momentum metaphor makes no justice at all to what actually happened. The project was not a case of a set of RRI specialists/activists who acted upon the regional authorities to achieve a sort of behavioural change in them. Indeed, similar to the collaboration in Lombardy (see Chapter 4), in Catalonia we also see a strong connection between the RRI researchers/consultants/experts partners (Science for Change and the Open Systems Group) and the administrative partner (the Generalitat and specifically its General Directorate of Economic Promotion and Regulation). The Generalitat played a very active role in defining the content of the project.

It is in this light that one can fully appreciate the quote we introduced earlier, said by the main administrative partner: "Projects don't change policies." Equipped with Occam's Razor and the timeline of TRANSFORM and the two RIS3CAT strategies, it may look like TRANSFORM was the efficient cause of the inclusion of RRI and CS into RIS3CAT2030. From the perspective of the Generalitat, however, TRANSFORM was a means by which an already desired strengthening of the regional strategy with the conceptual planes of RRI and CS could take place. It was rather a case of policy changing (or defining) the project than the other way around. Or better, it was a case of network governance where a diversity of actors were able to pursue common goals in the presence of slightly different long-term objectives.

There are interesting translations taking place in this instance of network governance. We have seen how the pilot activities were exemplars of a multifaceted translation of RRI as CS. Firstly, there is an element of gathering the needs and expectations of citizens in order to integrate them into policy- and decision-making processes. This idea is supplemented with the aim to work on "real challenges" (Int_01) in the region through "involving people that are (...) actively contributing to the research" (Int_01). Hence, actors on different levels are ascribed agency in these pilots, they are conceptualised as epistemic actors. Still, there was also an element of education and awareness raising in the translations of CS in the Catalan activities, which is quite common in CS projects, see for example, Strasser et al. (2019).

The insistence on real challenges and actual contributions is highly interesting from our point of view. The pilots imagined in the planning phase were left because the work in the think tank pointed towards something "real":

> I mean we talk first with the more active actors in the think tank and it was the city council of Mollet del Vallès on waste and they agreed to lead the challenge to, to define the pilot on that area. Then with Hospital de Sant Pau where we are working with endometriosis, and they just agreed. (...). So, we couldn't have the three but these two we're running and they're working very well because they are real.
>
> (Int_01)

Here we see how the idea of having "real" activities is closely linked to the model of CS as co-creation and a form of participatory governance that is a central pillar of this cluster. The think tank was the central means for integrating local administrative actors and other regional actors. What is important to note here, however, is that this kind of integration and creation of responsiveness may reach well beyond the project lifetime of TRANSFORM.

> And in fact, the idea at the beginning was to have much less people in the think tank. I think in the proposal you only need to have like ten people involved but because in this case, [NN] is the right key player and involved a lot of people. And then that's why we had so many people at the first session especially of the think tank and then the pilots were so successful.
>
> (Int_01)

There are two things that are noteworthy here. Firstly, this is about temporalities and the limits of R&I governance through project funding. One of the reasons why the think tank was considered successful by the project members is the fact that the regional administrative partners were able to draw on previous work in the selection of actors. Second, it was also this experience that led to a particular framing of the think tank:

> I told them that if we want to have impact we need that think tank. That was, it has been a like a process. For the think tank we selected stakeholders that were already somehow engaged in the work I was doing and that could have some relation to citizen science and we open it a little bit more also.
>
> (Int_03)

The think tank in this framing was a process to select not only pilot activities but also to build and stabilise relationships with actors from the Catalan R&I ecosystem. As such it built on previously established links and was a means to make use of those.

The reason why the Generalitat needed a project like TRANSFORM for this is – as one actor from the administration told us – that in order to do things differently, there is always a need for some sort of mandate: "it's an opportunity also to start talking" (interview). A project like TRANSFORM is described as an opportunity, or as an "umbrella," that can nurture this type of work within the administration. And this nurturing – doing something "different" – reaches beyond the administration and includes understandings and translations of CS of the different partners in the pilot activities.

The fostering and nurturing of the relationships depend, however, on keeping them interested while keeping them on track. This is done not by offering easy money or other resources, but rather by insisting on the underlying cause – whether it be formulated as transformative change, RRI, or citizen empowerment. And it is in this context that the insistence on challenges being "real" makes sense: "Real" here means meaningful for the local actors in their local context and, in the words of one informant, that it connects "with something that is already happening in the territory."

We shall later discuss the enormous contrast between the TRANSFORM experience and the kind of RRI work that happens in its original context, namely that of biotechnology and converging sciences and technologies (see Chapter 8). Before the introduction of the acronym RRI, the basic challenge of public engagement with what is typically thought of as technoscience is to anchor it in people's lived context. Projects such as TRANSFORM indicate a very different trajectory for the RRI concept, namely that of anchoring it in places and life-worlds, in the concerns of citizens whose health care needs are not met, and of public servants who worry about the insufficient waste separation in their municipality. Without this anchor, no actual transformation will take place; this is at least the hypothesis.

The transformations may look small from the outside: The doctors and the endometriosis patients at Hospital Sant Pau begin to talk to each other in a different way. At the city hall of Mollet, technicians and other civil servants establish a new form of dialogue. Students, school pupils, family members, and neighbours begin to talk together about recycling of waste. The sum of these place-based, local activities contributes to the overall momentum of RIS3CAT's mission and to the visibility and reputation of CS in Catalonia because they generate new witnesses and ambassadors and extend the networks.

It is a truism among researchers with a lot of experience with EU-funded projects that in EU projects, people "do what they do." One may have to promise novelty and radical departure from one's business as usual. However, at the level of practice, where there are often strict limits on time and money, most partners in such projects will play it safe and do what they are best at and have done many times before. What we are presenting in this chapter may be seen as an argument in favour of, if not conservatism, at least the value of maintenance, continuity, and care work in a time where the rhetoric of novelty and disruption is strong. The TRANSFORM project as it was conducted was not framed as something completely new or as something that is expected to initiate something entirely novel. Rather, it was part of a long lineage of activities. It is an enabler of sorts, which makes it possible for actors in the Catalan R&I ecosystem to try to do "something more." To do so, they crucially depend on work that is done by the project team before and after the actual project. One colleague from an administrative partner stated that "it never works if you build things from nowhere."

The lines of continuity, from before and well after the project, are also what matters when the individual partners are to assess what they gained from the project. In our conversations in the basement, the expression "it is more than that" emerged again and again. For Science for Change, TRANSFORM was more than the pilots and more than achieving the project objectives, in the sense that it opened up new arenas and ventures for CS in the future. For the

Generalitat, TRANSFORM was one of many pieces in its puzzle, another opportunity tailored to continue on the larger mission of transforming the Catalan R&I ecosystem and indeed improve Catalan society. And for the Open Systems Group, it was both of the above – but it was also another small experiment in real time to simultaneously study and take part in what happens when people's agency is strengthened in unconventional ways, in an open social system where ultimately the demarcation between "science" and "society" becomes less of an issue.

Notes

1 RIS3CAT 2014-2020, l'Estratègia per a l'especialització intel·ligent de Catalunya del període 2014–2020. https://fonseuropeus.gencat.cat/ca/ris3cat/2014-2020/. Accessed April 6, 2023.
2 https://povesham.wordpress.com/2014/09/10/citizen-science-in-oxford-english-dictionary/
3 https://www.barcelona.cat/barcelonaciencia/ca/ciencia-la-ciutat/la-ciencia-i-la-ciutadania/ciencia-ciutadana/oficina-de-ciencia-ciutadana
4 European Commission. 2012. Guide to Research and Innovation Strategies for Smart Specialisations (RIS 3). https://s3platform.jrc.ec.europa.eu/ris3-guide. Accessed April 6, 2023.

References

Arnstein, Sherry R. 1969. "A Ladder of Citizen Participation." *Journal of the American Institute of Planners* 35 (4): 216–24. https://doi.org/10.1080/01944366908977225.
Coenen, Lars, and Kevin Morgan. 2020. "Evolving Geographies of Innovation: Existing Paradigms, Critiques and Possible Alternatives." *Norsk Geografisk Tidsskrift* 74 (1): 13–24. https://doi.org/10.1080/00291951.2019.1692065.
European Commission. 2012. "Guide to Research and Innovation Strategies for Smart Specialisations (RIS 3)." Publications Office.
Fitjar, Rune Dahl, Paul Benneworth, and Bjørn Terje Asheim. 2019. "Towards Regional Responsible Research and Innovation? Integrating RRI and RIS3 in European Innovation Policy." *Science and Public Policy* 46 (5): 772–83. https://doi.org/10.1093/scipol/scz029.
Generalitat de Catalunya. 2014. "RIS3CAT: Research and Innovation Strategy for the Smart Specialisation of Catalonia." https://fonseuropeus.gencat.cat/web/.content/ris3cat/documents/angles/estrategia-ris3cat-en.pdf.
Generalitat de Catalunya. 2022. "RIS3CAT 2030: Estratègia per a l'especialització intel·ligent de Catalunya 2030." https://fonseuropeus.gencat.cat/web/.content/ris3cat/documents/2030/ris3cat-2030.pdf.
Gibbons, Michael, Camille Limoges, Helga Nowotny, Simon Schwartzman, Peter Scott, and Martin Trow. 1994. *The New Production of Knowledge: The Dynamics of Science and Research in Contemporary Societies*. London, Thousand Oaks, New Delhi: SAGE Publications. https://doi.org/10.4135/9781446221853.
Irwin, Alan. 1995. *Citizen Science, A Study of People, Expertise and Sustainable Development*. London and New York: Routledge.

_____. 2001. "Constructing the Scientific Citizen: Science and Democracy in the Biosciences." *Public Understanding of Science* 10 (1): 1–18.

Kullenberg, Christopher, and Dick Kasperowski. 2016. "What Is Citizen Science? – A Scientometric Meta-Analysis." *PLoS ONE* 11 (1). https://doi.org/10.1371/journal.pone.0147152.

Salas Seoane, Nora, Diana Reinoso Botsho, and Carla Perucca Iannitelli. 2022. "First-person Experiences of Endometriosis in Catalonia: Recommendations of Women with Endometriosis for Improving Healthcare Services and Public Policies." *TRANSFORM Policy Brief 2022*. https://doi.org/10.5281/zenodo.7248108.

Strasser, Bruno J., Jérôme Baudry, Dana Mahr, Gabriela Sanchez, and Elise Tancoigne. 2019. "'Citizen Science'? Rethinking Science and Public Participation." *Science and Technology Studies* 32 (2): 52–76. https://doi.org/10.23987/sts.60425.

Völker, Thomas, and Ângela Guimarães Pereira. 2023. "What Was That Word? It's Part of Ensuring Its Future Existence" Exploring Engagement Collectives at the European Commission's Joint Research Centre. *Science Technology and Human Values* 48 (2) 428–53. https://doi.org/10.1177/01622439211046049.

6 Co-design and co-creation for social innovation in the Brussels-Capital Region

Stories of co-design artefacts and glass-walled offices

The first impression before even entering the actual office space of Be Participation, a non-profit association located in a highly populated neighbourhood in the centre of Brussels, is the sheer number of various installations, apparatuses, and other artefacts in the entry hall. Traces of past engagements and materialised plans of issues to be discussed and debates to come. There is a clear Do-It-Yourself feeling to the Be Participation premises. Once inside, this impression only becomes stronger as one sees an open space with tables, white boards, some desks here and there, pens, sheets of paper, and easily movable furniture like chairs and desks. This is clearly a space that signals the aim to be flexible and adaptive to different publics-in-the-making, a space that rejects being limited to one particular *Anordnung*, Martina Löw's notion for the social orderings that are manifest in the material set-up of spaces (Löw 2001). As such, the material set-up affords certain activities and ways of interacting (Gibson 1977).

As such the Be Participation offices (Figure 6.1 (a) and (b)) are clearly different from what we experienced in Lombardy when entering the history-laden halls of Fondazione Giannino Bassetti in the centre of Milan with its walls of books and expensive wooden furniture which embody the longstanding collaboration with the regional administration and the strong ties between FGB and the Lombardian research and innovation (R&I) policy-making elites. Instead, the Be Participation space speaks to a DIY community, bottom-up initiatives, and different forms of multi-stakeholder engagement. They are also quite different from the offices and meeting rooms at INNOVIRIS, which one of the authors of this book (Völker) visited the day after spending time with our colleagues from Be Participation. The INNOVIRIS offices in the centre of Brussels – at the time of the visit mostly empty due to the ongoing COVID-19 pandemic – are set up according to principles of functionality typical for modern office buildings. Offices and meeting rooms separated by transparent glass walls characterise this building. This is where Regional Innovation Plans and strategies are being developed and funding decisions are made.

DOI: 10.4324/9781003371229-6

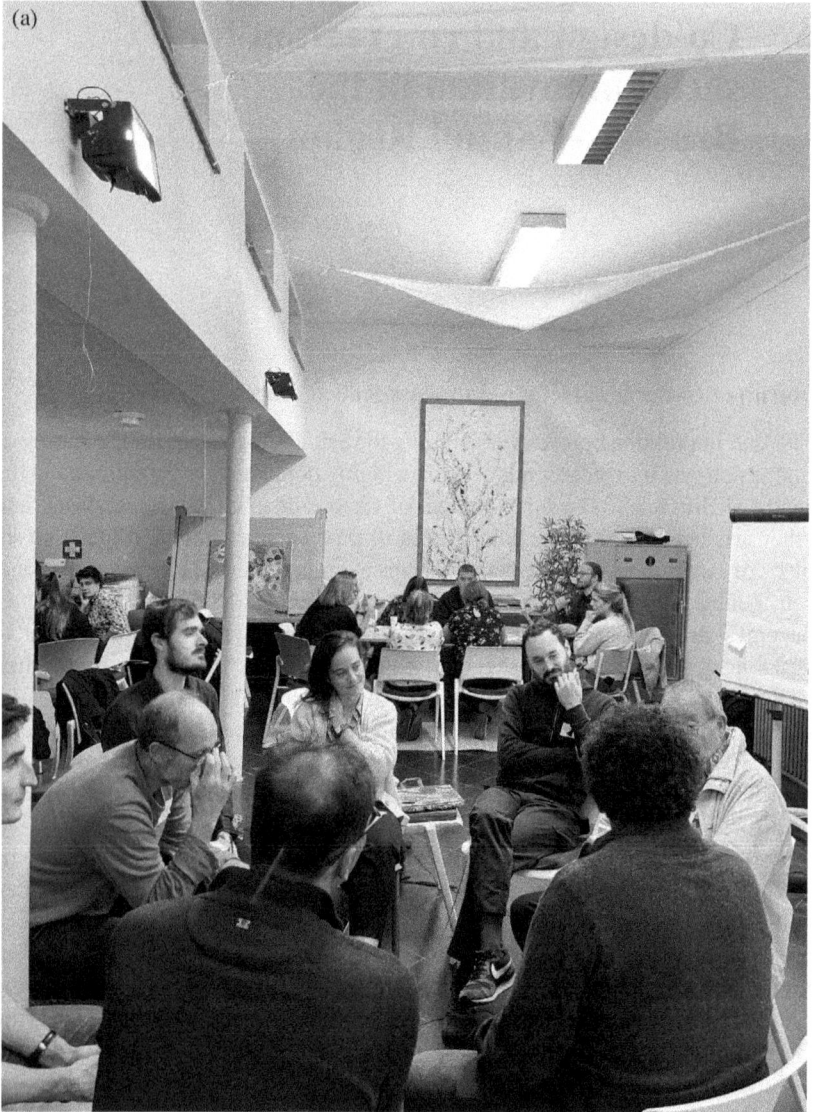

Figure 6.1 (a) Groups discussing in a participatory activity in the Be Participation offices. (*Continued*)

Figure 6.1 (Continued) (b) Additional group discussing in the Be Participation offices. Pictures taken by Marzia Mazzonetto from Be Participation.

This is the point of departure from which our colleagues in the Brussels-Capital Region (BCR) cluster[1] design and conduct their activities and translate Responsible Research and Innovation (RRI) into a dynamic and ever-changing R&I ecosystem.

Co-design and co-creation for social innovation – the BCR cluster of TRANSFORM

One of the core tenets of the BCR is that its governance structures consist of multiple layers and distributed responsibilities. Our interviewees describe the R&I ecosystem as "complex" (Int_11) and would often express the need to constantly re-evaluate the situation and their own position within it. The S3 platform describes this as "a small knowledge-intensive city-region, nexus of financial, geopolitical and multicultural flows and centre of the European Union."[2]

This situation makes for a fascinating task for this cluster, namely, to contribute to the "development of territorial RRI," which is a central aim of the SwafS-14 call that provides the funding of the TRANSFORM project. More than in the other clusters this task included mapping out what and who the *territory* actually is. In the case of the project's BCR cluster, this was to be done through collaborating with the regional R&I funding agency INNOVIRIS and Université catholique de Louvain.

The overall aims of this cluster – in project jargon it was referred to as a work package, "WP 5 - Design for Social Innovation – Brussels-Capital Region" – included the following elements as laid out in the TRANSFORM grant agreement:

- Set up and carry out a multi-stakeholder engagement process through design thinking approach.
- Address concrete community needs, emerging from a bottom-up participatory process involving at least 200 citizens and carried out in two distinct Brussels districts.
- Engage citizens, local CSOs, and universities/SMEs in the co-creation of social innovation solutions in the field of circular economy, leveraging on the urban metabolism of Brussels: Identifying specific resources to be reinjected into the local economy rather than exported as waste, supply chains and local economic actors to implement solutions, and market demand to benefit and scale up the proposed solutions.
- Foster novel and transparent governance relations within the regional R&I agenda setting.

Given the context of this project is the SwafS-14 call, these aims are hardly surprising. They are part of what needs to be offered to satisfy the expectations of this call. Still, it is important to reflect on the ambitions of a project like TRANSFORM (we have already highlighted this in previous chapters) and how these ambitions and objectives translate into a given R&I ecosystem.

In the case of the BCR cluster we are talking about a highly developed and complex ecosystem. BCR is home to around 1.2 million inhabitants and has an average population density of over seven thousand people per km². This means that this as a densely populated area. In addition, the region consists of 19 municipalities.

This is described as a "patchwork" and coming with "complex institutional challenges."[3] We will come back to what this means for the work of our colleagues in this cluster in the empirical parts of this chapter.

BCR is also the home to an active R&I ecosystem, bringing together five universities (including three of the seven Belgian university hospitals) and around 25 higher education institutions with a culture of tech-start-ups and key innovation actors of the European Union.

At the time of the TRANSFORM activities, the BCR was working on its S3 strategy for the period 2021–2027. INNOVIRIS, as the agency responsible for the process, set up a Steering Committee consisting of stakeholders from different sectors such as construction, finance, environment, and universities. This committee was tasked with defining societal challenges and mapping the local R&I ecosystem using the Entrepreneurial Discovery Process (EDP). At the time of our fieldwork, six challenges had been identified. These are (1) climate and energy, (2) resources optimisation, (3) inclusive and representative society, (4) health and well-being, (5) healthy food for all, and (6) mobility (see TRANSFORM Deliverable 5.1 for more information on this process).

In the new Regional Innovation Plan (PRI), which includes the renewed RIS3, these challenges correspond to five thematic strategic innovation domains (SIDs): (1) Climate: resilient buildings and infrastructures; (2) optimal resource use (this includes circular economy, which plays an important role in the work of the cluster); (3) efficient and sustainable urban flows for inclusive urban space management; (4) health & personalised and integrated care; and finally (5) social innovation, public innovation, and social inclusion. These SIDs are accompanied by a transversal theme of advanced digital technologies and services.[4] As such, the new RIS3 is described as a "change of perspective" that aims to take on board recent policy developments as exemplified in the European Green Deal (EDG).[5] Most of these domains focus on technoscientific innovation whereas domain five promises reflection of governance structures and forms of participation in policy- and decision-making processes.

These current activities build on a long history of innovation strategies in the region. Already back in 2006, a smart specialisation strategy was adopted as part of the PRI. This then was adopted by the regional operational plan of the European Regional Development Fund (ERDF).

The work of this cluster is situated within these thematic areas with its pilot activities geared towards food waste and the circular economy. The circular economy has been a topic within the region's innovation strategy for quite some time now. The BCR government adopted a Regional Program in Circular Economy in 2016, aiming to position itself as a European frontrunner in the area.[6] The circular economy thus also became a key priority for the region, with food as one of its focus areas.

RRI gets translated into this R&I ecosystem thus as a particular way of supporting these innovation strategy areas. In addition, the BCR cluster focuses on co-design and co-creation as its main methodological frames. Before we get into the thick of the stories we want to tell about the work of the BCR cluster, we need to devote some space to talk about what design thinking is and what it means to think about RRI in terms of design thinking, co-design, and co-creation.

Design thinking and multi-stakeholder engagement

One of the central pillars of the work of the BCR cluster is design thinking. The practice of designing is mostly associated with making technologies or products aesthetically pleasing and thus more attractive to users and consumers. Towards the end of the last century, design thinking became prominent within management debates, business circles, and in the consulting community. People like IBM's Thomas Watson Jr. or Apple's Steve Jobs have referred to design thinking when describing their work. Design thinking is also institutionally well-established, for example, in the d.school at Stanford University or in the School for Design Thinking at the Hasso Platter Institute in Potsdam (Carlgren, Rauth, and Elmquist 2016). In an article in the Harvard Business Review from 2008 Tim Brown – chair and CEO of IDEO, one of the main proponents of design thinking – describes design thinking as follows:

> Put simply, it is a discipline that uses the designer's sensibility and methods to match people's needs with what is technologically feasible and what a viable business strategy can convert into customer value and market opportunity.
>
> (Brown 2008, 2)

This description shows that in the view of its proponents, there is more to design thinking than its aesthetic dimension. It is about considering consumer-needs and integrating them into product design. Proponents of design thinking thus consider themselves problem solvers that are not only useful in making things aesthetically nice at the end of the production process, but rather as a part of the innovation process that should be present from the very beginning.

Famously – or perhaps also infamously – design thinkers distinguish different phases. While Brown and Katz (2011) initially talk about "three spaces of innovation" – inspiration, ideation, and implementation – these spaces are later developed into five modes: The empathise mode, the define mode, the ideate mode, the prototype mode, and the test mode. The empathise mode includes research and data collection with the aim to understand the user's perspective and experience. The define mode builds on that research and is about the creation of a shared problem framing including the desired outcomes and potential constraints. The ideate mode focuses on developing solutions for that problem, preferably as many solutions as possible so that there is material to experiment with. Finally, the prototype and test modes focus on producing tangible solutions that are tested with potential users. The crucial thing for design thinkers here is to produce something tangible, artefacts that can be tested and toyed with (Wolcott and McLaughlin 2020; Wolcott et al. 2021). Often, these designs include so-called "sacrificial concepts," early ways of capturing user experiences that later can be abandoned easily. The idea is to elicit strong reactions from the users to spark debate and potentially (supplemental or alternative) ideas.

More recently, Jon Kolko in his article "Design thinking comes of age" (Kolko 2015) argued that design thinking has moved beyond being a method for product design and is on its way to develop into a set of principles that can shape corporate cultures. The core element in this process is an organisation-wide focus on user experience and emotions, a reliance on creating models and artefacts to examine complex problems, the use of prototypes to develop potential solutions, a tolerance of failure, and a willingness to nurture a culture that is based on iterative processes. Importantly, a key pillar in design thinking as a culture is to exhibit restraint in the sense of focusing on the creation of a simple emotional value of certain products. Some proponents of design thinking argue that the focus on consumer experiences needs to be supplemented with an increased attention towards the needs and experiences of communities (Meroni and Sangiorgi 2011; Meroni and Fass 2013). In doing so the principles of design thinking are brought together with participatory approaches and action research for community development.

Over the years, the ideas expressed by proponents of design thinking have been criticised among designers, innovation scholars, and engagement practitioners. Sometimes these criticisms are formulated in quite drastic ways. Historian of science Lee Vinsel, for example, published a text with the title "Design thinking is kind of like syphilis — it's contagious and rots your brains."[7] Another article states: "Design Thinking is Bullsh*t."[8]

There are two main lines of criticism that go something like this: Firstly, what is often described as design thinking is an assemblage of vague terms without any clear meaning or pointers towards actual practice. It is a fad made up of buzzwords that can be filled with meaning and don't offer anything new beyond what people already know and like. As Vinsel describes it, "floating balloons of jargon, full of hot air." The second line of critique in a sense defends this vagueness of the terms and points to the iterative and experimental nature of design thinking and argues that it is the simplification of design thinking into methodologies, into linear step-by-step guides that is the problem (in order to package it into sellable products). What was intended to be an iterative process of tinkering and experimenting has been streamlined into step-by-step guides that are premised on linear processes. This is visible in papers that propose a series of steps or a number of tips to stimulate design thinking processes (Wolcott et al. 2021). The problem then is not so much the approach itself, but its marketing as a silver bullet solution to all problems.

This is also captured in a definition of design thinking, which Natasha Jen proposed in her talk "Design Thinking is Bullsh*t":

> Design Thinking packages a designer's way of working for a non-design audience by way of codifying design's processes into a prescriptive, step-by-step approach to creative problem solving — claiming that it can be applied by anyone to any problem.
>
> (Natasha Jen)

The solution for people within the design thinking community is to resist the urge to create an overly simplistic version of design thinking for the sake of popularity, and instead focus on the experimental and iterative nature of the design thinking process. This, however, leaves open the question of what is actually new about it and where the added value of design thinking actually lies. Despite this criticism, design thinking is still widely used also beyond the United States in Europe as visible, for example, in the work of Politecnico di Milano's School of Design[9] or the Design for Social Innovation towards Sustainability network (DESIS).[10]

The BCR cluster, in addition to making use of design thinking concepts and tools, also uses a version of multi-stakeholder engagement. In doing so, our colleagues draw on work promoting the inclusion of actors from the so-called "quadruple helix" (Leydesdorff 2012). Basically, the quadruple-helix model is a way to describe innovation governance as the interaction between actors from the university-industry-government complex, and the public. A quintuple helix has even been suggested, where the environment is included as an actor. These concepts were developed by Carayannis and Campbell (2010) and built on the initial work of Etzkowitz and Leydesdorff (1998).

Thus, while the inflation of helices may sound somewhat abstract, cerebral, or even off-putting, there is actually a genealogy behind it that reaches back to the closing decades of the previous century. These are debates about changing modes of producing and circulating knowledge, captured in terms such as "Mode 2" or "post-normal science." In this context, the notion of a "triple helix" was developed (Etzkowitz and Leydesdorff 1998) and subsequently attracted interest within the ongoing discussions on changing relations between (academic) science, administrations, industry, and public or civil society. Initially, the triple helix model was developed as a heuristic for analysing changing relations between universities, whereas Mode 2 (and probably to a lesser extent post-normal science) can also be seen as a diagnosis about ongoing developments. As a heuristic, it aims to direct attention to the relations between universities, industries, and governmental actors in innovation processes. The starting point, however, is the assumption that a transition is occurring that renders university-industry-government relations increasingly important for contemporary knowledge societies (Leydesdorff and Etzkowitz 2001). Thus, while one of the core interests of Etzkowitz and Leydesdorff is indeed changing ways of producing and circulating knowledge, they very much bring in an institutional perspective. Innovation scholars have therefore noted that their work "is focused on the organizational context of Mode 2 research" (Hellström and Jacob 2000, 76).

But why use the analogy of a helix? The reason for this is actually quite simple: By using the picture of a helix they want to highlight processes of iterative interactions between universities, industries and government actors and the production and governance of innovation. In addition, the metaphor of a spiral movement points to an element of reflexivity. Talking about a helix can thus be understood as a critique of simplistic linear models of innovation.

What makes this relevant for work within RRI pilot projects is that in discussions about the helices of different actors, participatory and collaborative ways of working play an important role. Where the post-normal science debate argues for engaging an "extended peer community" that includes people who are potentially affected or even negatively impacted by certain technoscientific innovations or policy decisions, Helga Nowotny – one of the main proponents of the Mode 2 literature – argues that so-called "transdisciplinary" modes of knowledge production can lead to more "socially robust knowledge" (Nowotny 2003). This interest in quality and robustness goes together with a call for "accountability" of science and innovation, which is also a crucial element of RRI.

> Social accountability permeates the whole knowledge production process. It is reflected not only in interpretation, and diffusion of results but in the definition of the problem and the setting of research priorities, as well.
>
> (Gibbons 1994, 62)

The triple-, quadruple-, and quintuple-helix models therefore share a certain concern with debates about Mode 2 and post-normal science, which is the ambition to initiate epistemic, social, political, and also moral re-orderings. In some of the scholarly literature from the last decade, these concerns are captured in notions such as responsibility and care (Mol 2008; Puig de la Bellacasa 2011, 2017; Arora et al. 2020).

In what follows, we will carve out the particular translation of RRI into the BCR R&I ecosystem and into approaches guided by design-thinking and quadruple-helix multi-stakeholder engagement.

Translating co-creation and design thinking

The BCR cluster conducted two pilot activities in which RRI was translated in two quite distinct ways: First, as an urban development project that sets out to deal with the problem of unsold food, and second as quadruple-helix engagements following a design thinking approach with young researchers at the Catholic University of Louvain.

Unsold food

The first pilot activity we want to talk about addressed the issue of unsold food. At the time of our fieldwork, several different initiatives were tackling the challenge of food waste in Brussels. These initiatives were working in parallel and in some cases even in competition with each other. The cluster partners Be Participation and INNOVIRIS organised engagement activities to co-design a process for solving the tensions created by this situation. The general aim of RRI, as it was translated in the BCR cluster's work, was to provide a "service" (Int_10) or to give "support" (Int_10) to publics already formed around a certain issue. The work on this pilot activity started out with an initiative called "No Javel!" which roughly translates as "no bleach"[11]:

> It's this project called No Javel! that is a citizen initiative, so it is completely unstructured. It exists as a not-for-profit thing, association, but it's totally handled by volunteers, citizens, and they go and get tons and tons and tons of unsold products from organic supermarkets, only organic places and they redistribute it to poor people.
>
> (Int_09)

This stance of dealing with already existing initiatives resonates with Callon and Rabeharisoa's idea of "the wild" (Callon and Rabeharisoa 2003) and with work by Noortje Marres on the simultaneous formation of publics and issues (Marres 2007). The pilot thus engaged with locally situated knowledge and lived experiences of food production and consumption systems, combined with social innovation for groups described as "disadvantaged" (Int_09) and as "people who don't feel entitled or interested" (Int_09) in participating in political processes. So, what we see is a clear normative stance in the objective to contribute to an improvement of regional governance approaches.

The ethos of *being of service* is crucial here: It is a central pillar of what it means to do good engagement in relation to the unsold food-issue in Brussels, as well as a translation of the idea of R&I becoming responsive to or aligned with society. In the unsold food pilot, this meant identifying needs and building networks – a version of responsibility that focuses on "facilitating meaningful engagement" (Int_09) with regional bottom-up initiatives focused on social innovation.

Simultaneously, the pilot aspires to be useful for the project partner from the regional administration. This particular idea of service that we see here – focused on the *sustainability* of the initiative – managed to integrate INNOVIRIS in the translation. For INNOVIRIS a central concern is the longevity of the initiatives they fund. This, then, is simultaneously about the initiative and maintenance of local RRI projects. Thus, the idea is a twofold one: On the one hand to help solve the problems by facilitating *meaningful engagement*, and on the other hand by finding ways to avoid such situations by influencing the evaluation grid at INNOVIRIS so that it is sensitive to these types of issues.

During the first half of 2022, the Brussels cluster planned and organised a series of workshops and other engagements in which a range of actors with stakes in the unsold food issue were invited to discuss and deliberate. The cluster organised four workshops with citizens and users of services or apps such as Too Good to Go, Phenix, or Happy Hours Market. They conducted group interviews with customers of a social grocery store, held a workshop with representatives of associations providing food aid from unsold food and another workshop with representatives of public authorities concerned by the flow of unsold food. In addition, they also did some individual interviews with representatives of Happy Hours Market, Too Good to Go, DREAM, Färm, Sequoia, and Bio c'Bon. Overall, they involved 50 citizens ("users"), six representatives of local and regional public authorities and funding bodies, more than 15 representatives of local associations (non-profit), innovation representatives, representatives of retailers and supermarkets and the private sector, and circular economy experts.

In October 2022, the cluster organised a sharing workshop to reflect on the common points identified and to start developing possible solutions, as well as proposals for possible next steps in the process.

For all of these activities, our colleagues tended to use the term "co-design" to point to the fact that the objective was not to actually solve all the issues with the unsold food situation in Brussels, but rather to map all the issues as perceived by the different actors involved and to co-design or collaboratively develop a process for solving them. We will come back to this later in the chapter. Before we do so, we want to briefly introduce the other pilot activities of the BCR cluster.

AquaSens and Algorella – working with innovators at UC Louvain

The second pilot of this cluster focused on practices of co-designing innovations, specifically on the development of water sensors (AquaSens) and in the broader area of the circular economy (Algorella). AquaSens is a technoscientific innovation – water sensors – developed by students from UC Louvain.[12] The aim of these sensors is to develop easy-to-use tools for testing water quality. The second UC Louvain student project was called "Algorella," which is basically a vegan pesto based on microalgae and nuts.[13]

In this pilot, the consortium partners Be Participation and the Catholic University of Louvain worked together with PhD students on their projects and innovations. RRI here takes on the meaning of design-thinking in the shape of quadruple helix workshops. RRI thus got translated into a network of Be

Participation (a civil society organisation), a university and its PhD committees, students and their innovation projects, selected actors from civil society, industry and academia, INNOVIRIS, and potential evaluation mechanisms at (potentially) several levels.

The quadruple helix workshops constituted a form of organised and guided deliberation. There were several interdependent and entwined aims which included (1) discussing the political issues involved in the innovations, (2) a process of collaborative prototyping, (3) feedback on the marketability of the innovation as a product, and finally (4) a showcase of the Spheres protocols (we will return to this in the next section).

These activities were explicitly discussed by the actors as a form of "upstream engagement" (Int_12) organised over a long span of time on several occasions with different foci. In them, the figure of *the innovator* is crucial. Innovators are developers of a certain product. This is obvious to the point of being trivial. But there is more than that. At the same time, these innovators are also PhD students primarily concerned with research. Their PhD research brings in a particular idea of innovation that resonates with RRI principles, which is the concept of a *360 degree view of innovation*. RRI translated as quadruple helix engagement here is explicitly linked to Jasanoff's idea of "technologies of humility" (Jasanoff 2003).

Interestingly, while the partners in this pilot argued strongly in favour of the integration of heterogenous actors into innovation processes, there was also a palpable attentiveness to the risks of such engagements and an awareness of the need to protect PhD students and their projects. We observed a careful demarcation or purification work on the side of the researchers. There are areas where engagement is "not interesting" (Int_12), which includes "highly technological" (Int_12) areas. In such areas, so our colleagues argue, also simple models of knowledge transfer still work. Thus, we see a simultaneous process of entanglement and purification. In addition, citizens appear in multiple roles: While they can provide valuable feedback that might improve the work of the innovators, they are also perceived as a potential risk to the projects. That is, there is a risk and potential for the projects to get in "trouble" (Int_12), for example, if the citizens negatively evaluate a PhD-project. We will return to that point later.

What we see in the BCR cluster thus are two distinct translations of RRI, one focusing very much on co-design and community-based needs and goals, the other centred around changing innovation cultures in the education of engineers at universities. Before we unpack these pilots further, we want to say a few words on the particularities of the BCR R&I ecosystem.

The BCR ecosystem – contingency and uncertainty

The R&I ecosystem and the position of the members of this cluster within it provide a contrast to the clusters in Lombardy and in Catalonia. While our colleagues from Lombardy and Catalonia are in a position where they can build on – and even further develop – pre-existing stable relationships with their administrative partners, the BCR cluster activities basically had to start from zero. The work of the BCR cluster is therefore and interesting case for asking what the absence of particular kinds of relationships means for RRI work and the impact that can be achieved through this work.

In the conversations we had with members of the BCR cluster, they described the governance structures and the overall R&I ecosystem as fragmented or "complex" (Int_11). Interestingly, we got descriptions like this by both researchers and the administrative partners. In addition to being complex and fragmented, the system was also described as being in "constant movement" (Int_09) or as a "cycle" (Int_09) of actors in decision-making positions. This situation, naturally, made it more difficult to assess who actually is a relevant actor:

> So, there's this constant movement of, because they are all functionaries, so civil servants. So basically, there were some of the people who were in the ministry ended up in Innoviris. Some of the people who were in Innoviris ended up in the in the government. So, there is constant cycle.
>
> (Int_09)

Given this constant movement, the main challenge according to our colleagues was to decode this R&I ecosystem, to identify the relevant actors to work with, and to build spaces for RRI-inspired pilot activities. In addition to that, also the issues to work on needed to be identified in a situation where topics and priorities tended to shift. Given this overall situation, it comes as no surprise that the activities in the BCR cluster are described as being more "disconnected" (Int_09) than the activities of the other clusters. These difficulties were already present in the set-up phase of the project and in its proposal stage. During that phase, different actors from the regional administration were considered as collaborators at different points in time. Only at the very end of this process, it became clear that INNOVIRIS would join the TRANS-FORM project. A colleague from INNOVIRIS described their participation in the project as "accidental" (Int_11). Another essential difference between the BCR cluster and the situation in the other regions is the standing of the research partners, who describe themselves as "newcomers" (Int_09) in the local R&I ecosystem.

The central idea of how to deal with this situation was to design a "protocol" for assessing innovation projects according to RRI principles. The assessment made on the basis of this protocol could then be used for advising these projects on how to become more resonant with RRI principles: This was the "Spheres protocol." In this protocol, RRI is translated as an assemblage of different *techniques* or *scripts*. Conceptualised like that it gains the ability to *travel* across different sites and sectors. It also can be used for capacity building among different R&I actors. This way of thinking about RRI necessarily involves an aim towards standardisation. This protocol – so the idea of the cluster partners – solves a number of problems that are very specific to this particular cluster: It can be theoretically applied in radically different settings. One such setting would the UC Louvain, where it can be tailored to introduce students to a broader view of innovation. It can also be adapted to work as an addition or extension of evaluation procedures at INNOVIRIS. The idea in this cluster was to develop a boundary object that would enable them to discuss and introduce alternative concepts of innovation through the evaluation infrastructures that are already in place.

At this point it is also important to mention that our colleagues referred to the Spheres pilot in different ways: They called it a *protocol*, a *prism*, a *service*, and also a *vehicle*. The common denominator of these different ways of describing this pilot is the intention to use it as both a lens for analysis and a set of guiding principles for giving input to R&I projects. This resonates with a framing of RRI as a service, in this case a service for researchers to take alternative approaches, methods, and perspectives into account – in accordance with RRI principles. There is also a strong element of research and analysis in this pilot. However, the aim is always to use RRI principles as a form of service to be provided, never as an end in itself.

Translating RRI in the BCR cluster pilot activities

After this brief outline of the different pilot activities, we will use the next part of this chapter to unpack the particular translation of RRI in these pilot projects. For doing that we will ask what the ideas of good engagement are that are implicit in the cluster's activities and then have a closer look at the translations of RRI that become visible in the various pilot activities and in the stories our colleagues tell about these activities.

Good engagement in the Brussels-Capital Region cluster

More than any of the other clusters, the BCR cluster is characterised by a diverse set of translations of RRI. RRI is translated as a co-design process to show possible ways of resolving a conflict about unsold food, a series of quadruple-helix engagements to improve PhD students' innovations together with their understanding of the concept of innovation itself, and as a way to improve the evaluation criteria of an R&I funding agency. As such the ideas about what RRI is range from community development to a particular approach to technoscientific research and development, and even to research evaluation.

We can get a first glimpse into these translations by looking at our colleagues' answers to the question "what constitutes 'good' engagement?" We didn't actually ask them this question directly, but many of our conversations during the lifetime of the TRANSFORM project were focused on what constitutes good engagement within a broader context of RRI. The ways in which our colleagues from the BCR cluster talk about good engagement differ in terms of the underlying models of engagement, the general aims and purposes, the subject positions available to different actors, and in regard to the position of the activities on the imagined continuum between RRI as community development or as a form of research. There are also interesting differences in how different project partners think about the impacts of their activities and the temporality of their activities.

One of the core ideas about good engagement in this cluster relates to the idea of *being of service*. RRI as co-design or quadruple-helix workshops are set-up as a form of providing a service to someone. In the case of the unsold food pilot, it is a service for underprivileged and vulnerable groups of people in Brussels who need support to get access to healthy food. It is also a service to the community to find better ways of dealing with food that is still edible but can no longer be sold in supermarkets. There is thus a strong component of community development. This is not RRI in the sense of taming technoscientific innovation processes. It is about social innovation that is very much grounded in the lived experience of local actors and communities. Engagement in this sense also addresses very *concrete issues*, but concrete in a very different way than what we saw in the Lombardy cluster for example.

In the case of the work of UC Louvain, the RRI pilot is a service for students. It is about helping students to get feedback on their projects. This can be simply a matter of improving product design and getting feedback from potential future users, but this service can also be more than that. In the view of the cluster members from UC Louvain, this service aims at improving the students' view on innovation itself. Broaden the view to include potential implications down the road and in this way educate future generations of engineers and innovators differently. This implies a deficit in how technoscientific innovation is usually understood (and governed) and thus clearly resonates with ideas about cultural shifts present in mainstream RRI discourses.

What is striking is that these different translations of RRI as engagement are mediated by the networks people build and by their position within these networks. RRI, as practised by an NGO, is – perhaps unsurprisingly so – very different from RRI as practised within a university or a funding agency. Exactly how, then, are they different? And what does this mean in practice?

As we already alluded to, the Brussels cluster conducted several different pilot activities with different – equally legitimate – translations of RRI. There is a collaboration on the issue of unsold food, work with PhD students from UC Louvain on water sensors and novel foods, work towards integration of RRI principles with INNOVIRIS, and lastly, the transversal work on the pilot called "Spheres."

In what follows we will describe the different translations of RRI that become visible through these pilot activities and the way in which our colleagues make sense of them. The overarching approach that was intended to guide these pilot activities was the so-called Spheres protocol, sometimes referred to by our colleagues from Brussels as the *pilot of pilots*. And that's where we want to start our story.

Spheres – the pilot of pilots

The Spheres pilot was – or was intended to be – the overarching methodology through which the different activities, partners, and versions of RRI were integrated. Spheres consists of a set of protocols that transverse the different groups within the cluster. RRI and engagement get translated as a methodology, a set of *techniques* or *scripts* that are both a process and a heuristic. As such, Spheres is conceptualised as a tool for analysing R&I projects from an RRI-inspired perspective and to provide the means needed to improve the RRI elements in such projects. Ideally, from the perspective of the BCR cluster partners, the Spheres protocols can become a tool that *travels* across different sites and sectors, a tool that can be taught from one actor to another. Therefore, it necessarily implies an objective of standardisation.

This is the *pilot of pilots* for the BCR cluster. The initial idea was that all the different activities of the cluster are guided by the Spheres protocol and are also used to experiment with it and improve the prototype. As we already mentioned, our colleagues described the Spheres pilot as a *protocol*, but also as a *prism*, *service*, or *vehicle*. As such it was intended to be used as a lens for analysis and as a set of guiding principles for possible inputs into these projects:

> [F]irst analysing what could be interesting to bring to the project from the RRI like tools and then organise some activities based on the need that we have identified to, yeah. To support like, and in the case of the water sensor project, for example, it was done through the organisation of workshops, citizen workshops.
>
> (Int_10)

Spheres thus resonates with the cluster's notion of RRI as a service. This service would help researchers to take into account issues they had not previously considered.

> And what is interesting is basically what we tried to also explain to the researcher is that you can develop a very nice innovation that will measure the quality of water, but if you don't pay attention to the perception that

citizens have, they will maybe say 'okay, fine, I can test the quality but I will still use the bottled water.' So, you have to understand if your innovation like replies also to a need or to so. Otherwise it can be completely disconnected from society and so yeah.

(Int_10)

While this description of the Spheres pilot clearly resonates with an ethos of RRI as providing a service, there is a strong element of re-thinking the relation between R&I on the one hand and society on the other. Similar to the work in the Lombardy cluster, society is thought of primarily in terms of *needs* to which innovations should respond.

The Spheres process that is alluded to in the quotes above broadly consists of three main steps: (1) The definition of the project problem and needs, (2) the creation of an action plan, (3) and the implementation of the action plan through a series of activities. This conceptualisation is where the influence of design-thinking becomes most clearly visible in that it promotes prototyping, and centres around the users of the innovations to come. The first phase of the Spheres protocol is akin to the empathise mode in design thinking. It is also here that citizens – in the form of quadruple-helix actors – should be involved in the definition of the problem (by expressing their needs) and the development of innovations. RRI enters the process as assessment criteria in relation to gender, ethics, open access, public engagement, and science education. Participatory processes and attention to these RRI dimensions were also envisioned in the development and implementation of the Spheres action plans. In the implementation phase, the role of the citizens was considered to be discussing potential ethical, social, and environmental issues and, in that way, validating particular ideas.

Given this use of the Spheres protocol as an overarching or transversal pilot, some of the activities were also committed to tinkering or experimenting with these protocols. In addition to attempts of mainstreaming principles of RRI, the Spheres protocols were showcased in order to prove the added value of this (mainly for the innovators). We observed this form of *piloting as showcasing* in the work of the other clusters.

At one point during the project, the Brussels cluster launched a call for project proposals using the Spheres approach. The idea was to find projects that could go through the process of being assessed and then consulted according to the principles outlined in the Spheres protocols. Unfortunately, there was not enough interest in this call for projects, so this process was stopped. Instead, our colleagues from Brussels themselves selected and contacted projects which they considered to be fitting pilot projects. This is also how the contact with "No Javel!" – which would then broaden and develop into the pilot project on unsold food – was first established.

Unsold food and the ethos of being of service

As we described in the introduction to this chapter, one of the main RRI activities in Brussels dealt with the issue of unsold food. This work was premised on a particular translation of RRI, with its own ideas about engagement, about the actors that need to be involved, and about the purposes and rationales of such an engagement.

Our colleagues in Brussels started this project by getting in touch with one of the initiatives dealing with the issue of unsold food in Brussels – "No Javel!" There are two main points that make this initiative a relevant actor for this kind of work: First, this initiative tackles issue that are relevant for citizens as it brings together environmental concerns with questions of social justice. Second, it was important for our colleagues that this initiative was created by the citizens themselves. It is "volunteers"(Int_09) who work on improving the ways in which unsold food is dealt with because it actually matters to them and their personal lives.

> So, basically what's interesting in this project is that completely bottom-up approach. So, it's a project that was created and that is managed completely by citizens. It's not that it's a project that is answering a call for projects or a specific funding. It's really an initiative coming from citizens.
>
> (Int_10)

RRI thus takes on elements of community development as our colleagues get involved with this initiative and present different possible ways of solving the problems that emerge when several actors with different priorities start working on the issue of unsold food.

Framed in this way, the general aim of RRI is to provide a service or to give support to publics already formed around a certain issue. This means that we are dealing with a form of uninvited participation (Wynne 2007). It also sits well within an ethos of post-normal science hat is concerned with extended peer communities (Funtowicz and Ravetz 1992, 1993). What these authors argue is that it is important to be attentive to how publics are formed and that citizen engagement projects can result in the creation of publics through their activities, publics that don't exist independently from them and that could to some extent be described as *artificial*. Unfortunately, this means that once the project is over and there is no more funding for their work, the publics might disappear. In that way, working with initiatives like the ones involved in the unsold food pilot project can be seen as a response to a major shortcoming of many engagement activities, which is that very often publics get formed top-down through engagement events or self-select due to socio-economic status (they have the education and spare time to do so). Importantly, working with initiatives that already exist prior to the engagement activities and thus tapping into what Chilvers and Kearnes (2015) call broader "engagement collectives" increases the chance of a life after the event for the groups and the issues they are working on. However, it is not solely the publics that are shaped by how citizen engagement is set up and organised – indeed, the issues are also shaped by this. When citizens are engaged and are working on an established initiative, they might already have formed their own understanding of the issue at hand. This understanding might be different from the understanding of the public authorities. While for INNOVIRIS this case is about sharpening their evaluation criteria so they identify the actor best-suited to address a particular problem in a sustainable way or to fill abstract circular economy policies with life, for the bottom-up initiatives involved here it is about getting supermarkets to give people in need access to still edible food that they would otherwise throw away. Importantly, engaging with bottom-up initiatives also makes sure that locally situated knowledge and the lived experiences of food production and consumption systems can become part of the formulation of policies.

These themes are also important for our colleagues who worked on these pilot projects. In the conversations we had, these groups of citizens were often referred to as *disadvantaged publics* and *people who don't feel entitled or interested* to participate in political processes.

> So I thought 'well, you know one of the biggest challenges that we have and the activity we do is to involve disadvantaged publics and people who don't feel entitled or interested for different reasons they don't.' So we thought 'okay, this could be a really nice collaboration.' So I told them 'look, what I can do on my side is bringing to the association activities and projects that are, can generate this meaningful engagement of the type of target audiences that you can mobilise around the association but it's going to be on the topics that are familiar to me. So it's going to be about science, technology and innovation.'
>
> (Int_09)

Of course, these are people who wouldn't usually sign up for a focus group on food production and consumption systems. What we see in this translation of RRI is a clear normative stance expressed in the objective to engage with people who have actual agency and clear interests. The reason for this is that our colleagues in the BCR cluster explicitly aimed at contributing to an improvement of regional governance approaches.

This ethos of *being of service* then is a central pillar of what it means to do *good engagement* in this pilot project. It also nicely summarises the underlying model of the relation between the project partners and the actors they engage with. This ethos expresses a version of the RRI idea where R&I systems are supposed to become (more) responsive to, or better aligned with, society. In the pilot project on unsold food, this meant engaging with pre-existing initiatives in order to make room for their framing of the issues at hand.

What we see here is therefore a version of responsibility that focuses on *facilitating meaningful engagement* with regional bottom-up initiatives focused on social innovation.

> And what was interesting for us is the fact that it's coming from citizens completely and we wanted to suggest to them this service that I was explaining before. What we decided to implement in the cluster here in Brussels to basically bring some representative from the different helices of the Quadruple-Helix and to reflect a bit on the governance maybe of this initiative to see how it could be sustained.
>
> (Int_10)

The particular idea of service that we see here focused on the *sustainability* of the initiative. This focus proved to be crucial for also integrating INNOVIRIS into this translation. For INNOVIRIS, a central concern in their work is the longevity of the initiatives they fund. The pilot activity, as a consequence, was simultaneously about the initiative and about the maintenance of local RRI projects in a more general sense.

One important feature of this translation of RRI and its particular model of engagement is a focus on a contentious issue. While lots of work in the context of deliberative democracy has the tendency to steer debates towards consensus (Felt and Fochler 2010; Turnhout et al. 2020), the issue of unsold food was characterised by increasing tensions between the different parties involved. While some of the actors were grounded in civil society, others had an economic interest in mind. All of the actors involved, however, expressed the ambition to contribute to the *public good*, be it by organising food for people in need or by helping to push the BCR towards a more circular future. However, in doing so the various actors clashed with each other, and instead of *co-benefits*, there turned out to be unforeseen negative consequences.

> So, you get food that is about to expire less expensive but this Happy Hour Market even has a specific aspect that is they tend to give it especially to poor people. (…) So, this has reduced their access to unsold products of good quality. So, they are in conflict with these things that are supposed to do good. In principle they are all against gaspillage, against throwing things away that are still edible and sometimes even social aspect of support in those in need. They've managed to be conflicting with each other on flux of certain products.
>
> (Int_09)

It thus comes as no surprise that the partners in this cluster were at times pushed into the position of a mediator, a position they refused to accept due to the limited capacities of this pilot (we will come back to the challenges that come with organising such engagements as *pilots*).

The idea, then, is a twofold one: On the one hand, it entails providing help with solving the tensions at hand. This is envisioned to be achieved by facilitating *meaningful engagement* and in doing so giving an example to the actors involved of how such a process can work. On the other hand, the pilot project is also intended to speak to INNOVIRIS in that it aims to develop ways such situations can be avoided in the future. One way to do this is by initiating discussions about re-framing the evaluation grid at INNOVIRIS in a way that makes it sensitive to such issues.

Thus, this translation of RRI with its particular model of engagement managed to enrol the administrative partner in this cluster because it resonated with practices that were already being implemented. In doing so, the pilot spoke to challenges that were relevant for INNOVIRIS. While it may not be in TRANS-FROM's project lifetime that these discussions become consequential at INNO-VIRIS, these issues were at least registered.

This model of engagement as support for bottom-up civil society initiatives is seen as an addition to a more prevalent model that translates engagement as *technology transfer*. One of the actors from the Brussels cluster shares her impression by talking about so-called *technology transfer offices*:

So, for example Innoviris has in all Brussels universities they have these so-called TTOs, so 'technology transfer offices'. And all these technology transfer offices within the universities have one person paid by Innoviris there that works there. So, they are really on the technology transfer side. So, in a way yes it's normal that they work with the private sector more. So of course collaboration of research pri/public-private collaboration for sure. But to me it's still a bit surprising that the, the civil society part is completely absent let's say because in the end lots of these applications whether they are then, they come to life in the Belgium or Brussels territory or anywhere else.

(Int_09)

From this perspective engagement is very much framed in terms of university-industry collaborations, with room for improvement when it comes to bringing in actors from civil society. Some actors working on the pilot activities of the Brussels cluster argue that re-accentuating the model of engagement also means moving away from a narrow focus on *acceptance* as a purpose of engagement:

And then he said that he used several times the word acceptation. So, he said that (…) this guy from Co-Create didn't think that this mainstreaming - in the end it comes back to the mainstreaming - could bring more acceptation of research outcomes and innovations. So, he said that he's particularly interested in this concept and this thing about acceptation and that he would like to go deeper into what changes but works better when some projects (…) like innovations like we did with these water sensors do some type of participatory co-creation RRI.

(Int_09)

Our colleagues engaged in the unsold food pilot clearly consider this a narrow understanding of acceptance and work towards moving beyond it. They disagree with stressing the importance of *acceptance*, because it explains the rejection of policies or technoscientific innovations in terms of a cognitive or information deficit. And indeed, such a stance points to important differences between enhancing the *acceptability* of an innovation by means of public debate about framings and (collectively imagined) ends and using engagement as a means for *manufacturing acceptance*.

After talking about the unsold food pilot, we will now turn to the other partners in this cluster and to their accounts of the activities they were involved in. This means moving on to INNOVIRIS and the Co-Create program and ask how this relates to translations of RRI in TRANSFORM.

INNOVIRIS and the Co-Create model

INNOVIRIS is the regional administrative partner of Be Participation and UC Louvain in the BCR cluster. It is a public organisation responsible for funding R&I on a regional level. As such INNOVIRIS occupies a distinct place within the regional R&I ecosystem that comes with a particular view on RRI, innovation governance, and models of engagement. And while we see clear synergies in the collaboration between the different partners in this pilot, there are nevertheless some tensions between the model of RRI as a *service* – as expressed in the unsold food pilot – and the work of INNOVIRIS.

From the perspective of INNOVIRIS, RRI is clearly an important idea. This idea, so we were told, is operationalised in one of their funding schemes named Co-Create. This program starts from contemporary societal challenges – even uses the term *crisis*[14] – and calls for applying co-creation approaches to increase societal *resilience*. This means involving actors who are *concerned* in the sense of being affected by these challenges.[15]

Similar to our other colleagues from the BCR cluster, INNOVIRIS thinks of this program as an update to the *technology transfer model*. Technology transfer is seen as an outdated concept that we need to move beyond. In a way, RRI is translated as a better version of technology transfer.

What is interesting is that in our conversations the name Co-Create was described as a misnomer, since what is done in the projects funded in this *special scheme* is better thought of as *co-research*. The term co-research describes two closely related ideas: First, there is the temporal argument that non-academic partners ought to be involved throughout the whole project duration. Second, there is the socio-epistemic idea that they should be conceived as actors capable of providing substantive input:

> But it's more co-research and what we expect is involvement in the re-search of all the partners. So not just saying one will do the action, one will do the implementation of the knowledge and the other one will gener-ate the knowledge but everybody in the project should be co-researcher and it must include also citizens, academics, administration maybe, non-profit organizations, enterprises or whatever they want. But all these peo-ple should be involved in the research and to generate knowledge from the experiences and from their own knowledge. Complex.
>
> (Int_11)

This conceptualisation should be understood as an explicit critique of the ideal of compartmentalisation in projects and with a particular temporality im-plied in the notion of *transfer*, when such activities are displaced to the end of a project. All of this, as our colleague from INNOVIRIS was keen to point out, should be happening throughout the project in a collaborative *travel*.

> And you have to do this, this travel all together and why is it so impor-tant, it's because and that's something we brought back a few weeks ago in TRANSFORM is how do we enter the knowledge transfer at the end of a participatory process. And in Co-Create it was also a question - and I discussed with my colleague a few weeks ago about that question - and for him you don't have any knowledge transfer at the end of Co-Create because everything is done during the research process for all the neces-sary participant and the rest would not work and that's a big problem. Knowledge transfer generally is quite difficult between university and the society. It's difficult.
>
> (Int_11)

What we see here is a rehearsal of one of the central ideas of RRI, which is that responsibility means upstream engagement, involvement of a broad range of actors early on in the research and development process. As a suc-cessor of ELSI/ELSA research, it was an explicit aim of RRI that discussions on ethical, legal, and social aspects of R&I should no longer merely be an afterthought but should rather be brought in right at the beginning of R&I pro-jects (Schomberg 2012; Strand 2019). Comments like these, however, are not just comments about RRI and co-creation. Additionally, they express a certain view about the quality of participation and engagement practices more gener-ally: Early involvement, empowerment, and agency have been key issues in this strand of literature from the very beginning (Arnstein 1969) and remain important criteria in the evaluation of citizen engagement activities (Wickson and Carew 2014).

Our colleague at INNOVIRIS is very aware of the perceived risks and dangers of such a conceptualisation of engagement, and that some actors are apprehensive about this.

> And also in Co-Create you give a certain power to the citizen to people that are not predictable. It's easy to deal with academic experts when you have the, when you organise a lot of juries. It's easy to deal with them. You know where they will be going, and you can predict what they will do and what they will decide. But with citizens we had a lot of surprises sometimes. You say, everybody says 'okay, this project is nice' and then you have the citizen expert that says 'I don't give a shit about your project. It will not help anybody. You don't know what's our reality.'
>
> (Int_11)

Quotes like this about the challenges and even counter-arguments to engaging citizens show a keen awareness of what is at stake here. It is indeed about a transfer of control and power. Naturally, those in power are not all too enthusiastic about it. Very clearly, our colleague states that thinking about engagement of citizens in this way means re-thinking notions of power and agency and to give up control to some extent. Becoming responsive to society, as RRI demands, introduces the possibility of unexpected outcomes. This will become an issue when we talk about the challenges of implementing RRI in doctoral education in the case of UC Louvain later in this chapter.

One way of dealing with *risks* like this is to be very careful about the actual practice of empowerment, which means being cautious about which kinds of power to give to which citizens. There is thus an element of what could be called "top-down bottom-upping."

> So, what could be also interesting for me - and I have not yet the vision - is how to find the citizens. That's maybe also a big challenge because if you want to avoid professionalisation of citizens, that's one thing. If you want to avoid finding the citizens in the same pool of what we call "politicized citizens" so people that are active in associations is one. I don't say it's bad thing that they are involved in but if you consult them, they have already an orientation in the policies they want to defend. So that's a difficulty for us, to find the right person.
>
> (Int_11)

This model of engagement thus comes with a set of subject positions, ideas about what roles the citizens can play, which repertoires of interaction are available to them, and which prior normative commitments they (are allowed to) bring into the engagement. This is described as the challenge of finding *the right* people. While it is difficult to describe who the right person might be, our colleague from INNOVIRIS here talks about *professionalised citizens* and *politicised citizens* as subject position that might be challenging for public administration actors. There are two main elements to this: These terms describe citizens who have pre-formed opinions and positions. These can relate to a particular policy or to a certain issue of contestation. In addition, there is some form of organisation in the sense that there is a collective of citizens that exists independently from the engagement activities.

In addition to the professionalised and politicised citizens, the figure of the *unconcerned citizen* figures prominently in the account of our INNOVIRIS colleague. This figure comes into play especially in the ex-ante evaluation of projects at INNOVIRIS. The argument is that it is a problem if the citizens who are evaluating project proposals are *not concerned*:

> "So, sometimes you have such a feedback from the citizens that tell you what do it's far from all consideration. 'You tell us ah, we will make that all together but I don't give a shit of your things because I'm not concerned about it.'

> (Int_11)

This conceptualisation of the role of citizens is in line with ideas of the importance of lived experience and so-called "real-world problems" from discourses on transdisciplinarity. It also resonates with debates in the literature about matters of care and concern (Latour 2004; Puig de la Bellacasa 2011) and extended peer communities in post-normal science (Funtowicz and Ravetz 1992, 1993). This is an interesting conceptualisation of citizens: Their involvement here is clearly not thought of merely as a source of data or embodiment of values. Rather this relates to the perceived quality of the research projects. Involving citizens in this way means moving beyond an idea of quality that focuses mainly on narrow ideas of excellence, and instead aims at including the citizens' ideas about relevance. When a project manages to meet the concerns of the citizens, this means there is some relevance. *Giving a shit* in this way becomes a quality criterion in terms of the local relevance of a project proposal.

Statements like this perform an additional translation: Global challenges and wicked problems are translated as local challenges in certain neighbourhoods. They need collaborative research so that the people who have the lived experience of these problems can actually contribute. Compared to the bottom-up character of the unsold food pilot project, the model of collaboration present in the Co-Create funding scheme doesn't go as far in transferring power from researchers and innovators to citizens or *disadvantaged publics*. Consequently, there is still a danger that projects funded in this scheme are not able to adequately represent the lived experiences of the people in the region of interest.

This also says something about their idea of what *meaningful* engagement looks like and what structural barriers are. Meaningful engagement follows *a logic of service or delivery* and moves away from *a logic of discovery*; thus it is premised on a distinct translation of science-society relations (Felt 2017). From this perspective, funding programs that subtly favour work done in a logic of discovery can be seen as a structural barrier to RRI work in the regions.

Working with student innovators at UC Louvain

The third partner in the BCR cluster, the Catholic University of Louvain's Earth and Life Institute, worked with PhD students in the broad area of circular economy. This pilot centres around practices of co-designing innovations in quadruple-helix workshops. We already talked about these innovations: Water sensors (AquaSens) and novel foods that aim at increasing the circularity of food production and consumption systems (Algorella).

This pilot represents another distinct line of work within this cluster and an additional translation of RRI. There are some similarities with what we described above, but also some key differences in how RRI is translated. These differences are connected to the particular site of the translation: Doctoral education within a university setting. The overarching question for the work lead by UC Louvain, then, was how to integrate RRI into the education of engineering PhD students? What are the potentials and what are the risks?

RRI in this pilot gets woven into a very particular network of actors that is clearly distinct from the other pilots in the overall TRANSFORM project. Firstly, we have Be Participation, the cluster leader organised in the form of an association. Then there is UC Louvain, a university founded in 1834 that currently hosts around 30,000 students[16] and considers itself to be the country's leading French-speaking university, as well as the most international Belgian university. To make things more interesting, there are also university PhD committees, selected actors from civil society, industry and academia, INNOVIRIS, and potential evaluation mechanisms at (potentially) several levels. And of course, there are the students and their innovation projects.

This combination of actors comes with a set of commitments and expectations, which lead to a rather unique version of RRI. This version is focused on design thinking as an overarching approach to engagement, the ambition to impart a so-called *360-degree view of innovation* on the students, a particular idea of humility and the desire to safely guide PhD students through their projects.

The engagements in this pilot were organised as so-called quadruple-helix workshops. These are basically stakeholder workshops in which innovations are introduced and discussed with members from academia, politics, industry, and (civil) society. As such this takes the form of an organised and guided deliberation. There were several workshops, but the basic structure can be illustrated by going through the workshop from the AquaSens project. The workshop addressed (1) the political issue of water use and plastic bottles in Brussels to contextualise the innovation at hand; it (2) contained a collaborative prototyping of a water sensor as well as (3) some input on the marketability of the sensor as a product; and (4) finally – and this is in the specific context of a SwafS project that needs an outcome in the form of some sort of *impact* – a showcasing of the *Spheres protocols*.

The issue of quality was very prominent in our conversations about these two projects, especially regarding the education of PhD students and how to think about the innovation process. Good engagement here is discussed explicitly as a form of *upstream engagement*. Different actors, in this view, need to be integrated early in the process. In addition, our colleagues talked about the importance of having such engagements or deliberations over a long period of time, and on several occasions with different foci.

One of the central figures in the accounts of our colleagues working on these projects was *the innovator*. Innovators are the developers of a certain product to be – in this case sensors to measure water quality or novel foods. At the same time, however, they are also PhD students devoted to their own research and to finishing their PhDs successfully. The research they do in their PhDs brings in a particular idea of innovation that resonates with RRI principles. The main concept here is the idea of a *360-degree view of innovation*. This is a way of talking about RRI and engagement which is explicitly linked to the concept of a "technology of humility" as developed by Sheila Jasanoff (2003, 2007). Jasanoff uses the notion of humility to challenge a particular model of the relationship between science and politics premised on the idea of *speaking truth to power*. What she calls humility means acknowledging the limits of scientific knowledge and knowing, or put differently "when to stop turning to science to solve problems" (Jasanoff 2007, 33). This stance is not to be confused with anti-scientism. Much to the contrary it is about being more reflexive about the role science should be playing when entering the field of politics. Some questions simply are not easily answered by science alone:

"In the case of climate change, for example, science cannot tell us where and when disaster will strike, how to allocate resources between prevention and mitigation, which activities to target first in reducing greenhouse gases, or whom to hold responsible for protecting the poor. How should policy-makers deal with these layers of ignorance?"

(Ibid)

Technologies of humility, then, are ways of dealing with uncertainty, ignorance, and indeterminacy by directing attention to framings, vulnerabilities, questions of distribution, and potentials for collective learning. One way of doing so is re-thinking modes of knowledge production and governance by integrating "extended peer communities" (Funtowicz and Ravetz 1992, 1993) into knowledge- and decision-making processes. These ideas not only resonate with discussions about RRI in the BCR cluster and its pilot activities, but they have actually informed the design of these activities.

Our colleagues working on these pilots, then, argue for an early and continuous integration of a diverse set of actors, with the aim to point to the questions that go beyond what a narrow focus on innovation as a mere technological artefact or product could unveil. There is a sense of *openness* to this approach, and this is what our colleagues from UC Louvain talk about when they refer to a 360-degree view of innovation. While this is fully compatible with an RRI ethos, it is crucial to mention that there is a palpable attentiveness to the *risks* of such engagements. These risks are mainly expressed as a sensitivity to the need to *protect PhD students* and their projects from an undue influence of other actors.

> So, we are really taking true research from the university. I am asking to some in that case it is two PhD students. I am in their PhD committee. I asked them to enter the process because I thought it was interesting for them, but we are taking a huge risk because we take true research, ongoing research and we enter them as example in our prototype, to test our prototype but that could fire back to those students. Ok. And that's, you can see one of the students is today in the newspaper because of his work but it could really cause a lot of trouble if the experience in all prototypes is going wrong because you are testing something.
>
> (Int_12)

In this quote a member of the project team, who is simultaneously part of the PhD committee of one of the innovators, explains the risks of enrolling them into this *prototype* or pilot activity. Actual *ongoing research*, so this colleague argues, becomes part of an experiment on how to actually do research. Doing so comes with a set of risks that *could really cause a lot of trouble.* What we see expressed here is the tension between an ambition to expose the PhD students as innovators to the views and opinions of citizens and stakeholders and the urge to be careful and protect the students and their project from potential harm. This protective drive then manifests in the form of *demarcation or purification work* on the side of the innovators. In their view, there are areas where engagement is not *interesting*. These are, for example, *highly technological* areas for which simple models of knowledge transfer provide a suitable framework of engagement. Work on the circular economy, we are told, needs input from the mining industry. Engaging experts from the mining industry in the project is then a simply question of technology transfer.

What happens here is that by referring to the figure of the *PhD student innovator*, and the perceived need to protect their work from harm, we can see simultaneous processes of entanglement and purification.

Citizens do appear in multiple roles in the account presented by the cluster members from UC Louvain: As *providers of valuable feedback* for the innovators that can help to improve their work and the final products. But importantly, as we just saw, citizens are also *troublemakers*. This idea resonates with discussions about how to integrate RRI principles and participatory approaches into the evaluation mechanisms at INNOVIRIS. Publics or citizens are understood here as a potential *hurdle or obstacle* in need of being *tamed*. From this perspective, the main risk is that citizens (or evaluative citizen panels) might have the power to stop a project by evaluating it negatively. To solve this (mostly fictional) issue our colleagues argue for multiple engagements with citizens during a project's lifetime. In this way, citizens get a more complete picture of what is going on in the research projects and can make informed decisions.

The critical question for the cluster members involved in this work was about the level of *maturity* at which innovations can reasonably – and safely – be judged? What we see in this account is a tension between upstream ambitions and the need to protect PhD students that can lead to a downstream push. How far up the stream can you safely go? The overarching question then becomes who should be granted the authority to make these decisions. Who should define spaces and issues of legitimate engagement?

Contingencies in the territory

What we see in the BCR cluster is a fascinating case that shows what happens in a cluster configuration, where the relations between the different partners are not as stable as in Lombardy and Catalonia. Our colleagues from this cluster described the situation in the R&I ecosystem as multi-layered and complex. Their own position in that field is that of *newcomers*, which is in particular true for Be Participation. In addition, there was only reluctant commitment from the administrative partners – *accidental* as they called it – for several reasons not all of which were completely within their control. However, this presented a contrast to the organisational configurations where there was for example a strong lead from the Generalitat de Catalunya.

As a consequence, this translation foregrounded the element of tinkering and experimenting in the sense of *proofing* or *showcasing* for an open and curious but not yet fully convinced administrative partner. This situation led to several shifts in the pilot activities and also some attempts that didn't develop as hoped.

What emerged in this cluster as a consequence of this overall situation are two distinct translations of RRI, one focusing primarily on co-creation and community-based needs and goals, the other centred around changing innovation cultures in the education of engineers at universities.

RRI in the version we saw in the unsold food pilot takes the shape of a co-design process for community development. The ethos of this activity is one of *being of service*. Naturally, the follow-up question here is "of service for whom?"

In the case of the unsold food pilot the answer was twofold: In the beginning, this was clearly intended as a service for a local initiative on unsold food which got into trouble as new competitors with clear economic interests entered the scene and made it more difficult for them to get access to food. As it turned out, this case was also relevant for their regional administrative partner as they had also stakes in the issue of unsold food and food waste more broadly.

In this way the SwafS-14 call and thus the TRANSFORM project aims could be integrated into this translation of RRI together with the local community or the regional administration as a representative of the territorial R&I ecosystem (with regard to its governance structures). The SwafS-14 call and thus also the TRANSFORM project focus very much on shifts within regional innovation cultures in terms of policies and strategies and this indeed was the initial plan also for this cluster. However, the way the project was embedded in the regional R&I ecosystem and the position of the different partners – Be Participation in particular – led to a tendency of being of service to the local community.

UC Louvain's answer to the question of *service* turned out to be multi-faceted. First and foremost, they considered the pilot activities as a service for their students. This became visible in stories about the need for *protection* and subtle *purification-work* amidst processes of engagement expressed in the accounts of the cluster members working on the AquaSens and Algore-lla projects. In addition, one could argue that the work within the TRANS-FORM project also serves a broader purpose which is presented as a service for the university. This service consisted in a showcase for how to improve the education of future engineers, innovators, and members of the BCR R&I ecosystem. Attempting to establish a *360 degree view of innovation* clearly shows an ambition to shift the regional R&I culture towards integrating RRI principles.

What we also see in the pilots in the BCR cluster are the limits of *piloting* or – to put it less drastically – the *challenges of pilotification*. Our colleagues in Brussels had to navigate the difficult situation of conducting an activity with local initiatives that did not have an actual mandate. Or to put it differently, the mandate was to provide a showcase for the added value of co-creation and co-design approaches guided by RRI principles. In this sense piloting is *a proof of concept*. The risk with such a framing of the cluster activities – and this is true to different extents for all the clusters – is that very easily expectations can be disappointed. As one of our colleagues told us, there was a constant push to become something like a mediator between different parties who wanted to work with unsold food. This is something that our colleagues clearly did not have the authority to do as they are not a government agency and were instead

part of a SwafS-14 project with certain impact expectations and a rather limited project duration. For that same reason, it was important for our colleagues
from this cluster to talk about *co-design* instead of *co-creation*. The reason for
this is that co-creation would imply *tangible results* for the different actors
who were part of the unsold food pilot in the sense of a solution to the conflict.
What the BCR cluster could offer instead was a *co-designed* process through
which the different parties involved might be able to arrive at solutions at
some point.

One danger here is of course to disappoint people who put lots of effort into
working on issues that are regionally relevant and in doing so dedicate their
spare time to these pilot activities. Participating in such a pilot that cannot actually solve these issues might turn out to be counterproductive unless clear
follow-ups are provided. Being caught in the limbo of *piloting and showcasing*
thus might actually hurt the ambitions of achieving a cultural shift towards more
RRI.

This is an interesting parallel to the institutional relations in the Lombardy
and Catalonia clusters, which pose similar yet distinct challenges when it comes
to the broader framings of their respective activities. The following chapter will
move beyond the activities of the single clusters and provide a comparative perspective of the different regional cases.

Notes

1 Unfortunately, we were not able to visit the third partner of this cluster, UC Louvain,
and had to be content with having conversations in online meetings.
2 https://s3platform.jrc.ec.europa.eu/en/w/designing-a-policy-mix-for-responsible-
research-and-innovation. Accessed February 21, 2023.
3 https://s3platform.jrc.ec.europa.eu/en/w/designing-a-policy-mix-for-responsible-
research-and-innovation. Accessed February 21, 2023.
4 https://innoviris.brussels/sites/default/files/documents/innoviris_plan_regional_
innovation_pri_digital_fr.pdf. Accessed February 21, 2023.
5 https://commission.europa.eu/strategy-and-policy/priorities-2019-2024/european-
green-deal_en. Accessed February 21, 2023.
6 https://www.circulareconomy.brussels/?lang=en. Accessed February 21, 2023.
7 https://sts-news.medium.com/design-thinking-is-kind-of-like-syphilis-its-cont
agious-and-rots-your-brains-842ed078af29. Accessed November 17, 2022.
8 https://99u.adobe.com/videos/55967/natasha-jen-design-thinking-is-bullshit. Accessed
November 17, 2022.
9 https://www.design.polimi.it/en/. Accessed November 17, 2022.
10 https://www.desisnetwork.org/. Accessed November 17, 2022.
11 The name comes from instances in which people looking for leftovers of supermarkets got hurt by chemicals in the food waste. https://nojavel.org/. Accessed February
25, 2023.
12 https://beparticipation.be/transform-pilote-aquasens/. Accessed February 25, 2023.
13 https://uclouvain.be/fr/etudier/actualites/trois-etudiants-uclouvain-en-sciences-
agronomiques-primes-pour-leur-pesto-innovant.html. Accessed February 25, 2023.
14 https://innoviris.brussels/program/co-creation. Accessed February 26, 2023.

15 https://innoviris.brussels/sites/default/files/documents/presentation_co-creation_
    2021.pdf. Accessed February 26, 2023.
16 https://uclouvain.be/en/discover/faits-et-chiffres.html. Accessed November 15, 2022.

References

Arnstein, Sherry R. 1969. "A Ladder of Citizen Participation." *Journal of the American Institute of Planners* 35 (4): 216–24.
Arora, Saurabh, Barbara Van Dyck, Divya Sharma, and Andy Stirling. 2020. "Control, Care, and Conviviality in the Politics of Technology for Sustainability." *Sustainability: Science, Practice, and Policy* 16 (1): 247–62. https://doi.org/10.1080/15487733.2020. 1816687.
Brown, Tim. 2008. "Design Thinking." *Harvard Business Review*, 2008. https://hbr.org/ 2008/06/design-thinking.
———, and Barry Katz. 2011. "Change by Design." *Journal of Product Innovation Management* 28 (3): 381–83. https://doi.org/10.1111/J.1540-5885.2011.00806.X.
Callon, Michel, and Vololona Rabeharisoa. 2003. "Research 'in the Wild' and the Shaping of New Social Identities." *Technology in Society* 25 (2): 193–204.
Carayannis, Elias G., and David F. J. Campbell. 2010. "Triple Helix, Quadruple Helix and Quintuple Helix and How Do Knowledge, Innovation, and Environment Relate to Each Other?" *International Journal of Social Ecology and Sustainable Development* 1 (1): 41–69. https://doi.org/10.4018/jsesd.2010010105.
Carlgren, Lisa, Ingo Rauth, and Maria Elmquist. 2016. "Framing Design Thinking: The Concept in Idea and Enactment." *Creativity and Innovation Management* 25 (1): 38–57. https://doi.org/10.1111/CAIM.12153.
Chilvers, Jason, and Matthew Kearnes. 2015. *Remaking Participation: Science, Environment and Emergent Publics*. Abingdon and New York: Routledge.
Etzkowitz, Henry, and Loet Leydesdorff. 1998. "The Endless Transition: A 'Triple Helix' of University – Industry – Government Relations." *Minerva* 36: 203–8.
Felt, Ulrike. 2017. "Under the Shadow of Time: Where Indicators and Academic Values Meet." *Engaging Science, Technology, and Society* 3: 53. https://doi.org/10.17351/ ests2017.109.
———, and Maximilian Fochler. 2010. "Machineries for Making Publics: Inscribing and De-Scribing Publics in Public Engagement." *Minerva* 48 (3): 219–38. https://doi. org/10.1007/s11024-010-9155-x.
Funtowicz, Silvio, and Jerome Ravetz. 1992. *"Three Types of Risk Assessment and the Emergence of Post-Normal Science."* In *Social Theories of Risk*, edited by Sheldon Krimsky and Dominic Golding, 251–73. Westport: Praeger.
———. 1993. "Science for the Post-Normal Age." *Futures* 25 (7): 739–55. https://doi. org/10.1016/0016-3287(93)90022-L.
Gibbons, Michael. 1994. "The Emergence of a New Mode of Knowledge Production." In *Social Studies of Science in an International Perspective. Proceedings of a Workshop, University of Vienna, 13-14 January 1994*, edited by Ulrike Felt and Helga Nowotny. Wien: University of Vienna.
Gibson, James J. 1977. "The Theory of Affordances." In *Perceiving, Acting, and Knowing: Toward and Ecological Psychology*, edited by R. Shaw and J. Bransford, 67–82. Hillsdale NJ: Erlbaum.

Hellström, Tomas, and Merle Jacob. 2000. "Scientification of Politic or Politicization of Science? Traditionalist Science Policy Discourse and Its Quarrels with Mode 2 Epistemology." *Social Epistemology* 14 (1): 69–77.

Jasanoff, Sheila. 2003. "Technologies of Humility: Citizen Participation in Governing Science." *Minerva* 41 (3): 223–44.

_____. 2007. "Technologies of Humility." *Nature* 450 (7166): 33.

Kolko, Jon. 2015. "Design Thinking Comes of Age." *Harvard Business Review* (September): 66–71.

Latour, Bruno. 2004. "Why Has Critique Run Out of Steam? From Matters of Fact to Matters of Concern." *Critical Inquiry* 30 (2): 225–48. https://doi.org/10.1086/421123.

Leydesdorff, Loet. 2012. "The Triple Helix, Quadruple Helix, ..., and an N-Tuple of Helices: Explanatory Models for Analyzing the Knowledge-Based Economy?" *Journal of the Knowledge Economy* 3 (1): 25–35. https://doi.org/10.1007/s13132-011-0049-4.

_____, and Henry Etzkowitz. 2001. "The Transformation of University-Industry-Government Relations." *Electronic Journal of Sociology* 5: 338–4.

Löw, Martina. 2001. *Raumsoziologie*. Frankfurt am Main: Suhrkamp Taschenbuch Wissenschaft.

Marres, Noortje. 2007. "The Issues Deserve More Credit: Pragmatist Contributions to the Study of Public Involvement in Controversy." *Social Studies of Science* 37 (5): 759–80. https://doi.org/10.1177/0306312706077367.

Meroni, Anna, and Davide Fassi. 2013. "Design for Social Innovation as a Form of Designing Activism. An Action Format." In *Nesta*, 1–15. www.sustainable-lifestyles.eu.

Meroni, Anna, and Daniela Sangiorgi. 2011. *Design for Services. Design for Services.* London: Routledge. https://doi.org/10.4324/9781315576657.

Mol, Anemarie. 2008. *The Logic of Care: Health and the Problem of Patient Choice.* Abingdon and New York: Routledge.

Nowotny, Helga. 2003. "Democratising Expertise and Socially Robust Knowledge." *Science and Public Policy* 30 (3): 151–56.

Puig de la Bellacasa, Maria. 2011. "Matters of Care in Technoscience: Assembling Neglected Things." *Social Studies of Science* 41 (1): 85–106.

_____. 2017. *Matters of Care: Speculative Ethics in More than Human Worlds. Matters of Care: Speculative Ethics in More than Human Worlds.* Minneapolis: University of Minnesota Press. https://doi.org/10.1080/14636778.2019.1586527.

Schomberg, René von. 2012. "Prospects for Technology Assessment in a Framework of Responsible Research and Innovation." In *Technikfolgen Abschätzen Lehren: Bildungspotenziale Transdisziplinärer Methoden*, 39–61. VS Verlag für Sozialwissenschaften. https://doi.org/10.1007/978-3-531-93468-6_2.

Strand, Roger. 2019. "Striving for Reflexive Science." *Fteval Journal for Research and Technology Policy Evaluation* 48: 56–61. https://doi.org/10.22163/fteval.2019.368.

Turnhout, Esther, Tamara Metze, Carina Wyborn, Nicole Klenk, and Elena Louder. 2020. "The Politics of Co-Production: Participation, Power, and Transformation." *Current Opinion in Environmental Sustainability.* https://doi.org/10.1016/j.cosust.2019.11.009.

Wickson, Fern, and Anna L. Carew. 2014. "Quality Criteria and Indicators for Responsible Research and Innovation: Learning from Transdisciplinarity." *Journal of Responsible Innovation* 1 (3): 254–73. https://doi.org/10.1080/23299460.2014.963004.

Wolcott, Michael D., and Jacqueline E. McLaughlin. 2020. "Promoting Creative Problem-Solving in Schools of Pharmacy with the Use of Design Thinking." *American Journal of Pharmaceutical Education* 84 (10): 1271–76. https://doi.org/10.5688/ajpe8065.

——, Devin K. Hubbard, Traci R. Rider, and Kelly Umstead. 2021. "Twelve Tips to Stimulate Creative Problem-Solving with Design Thinking." *Medical Teacher* 43 (5): 501–8. https://doi.org/10.1080/0142159X.2020.1807483.

Wynne, Brian. 2007. "Public Participation in Science and Technology: Performing and Obscuring a Political–Conceptual Category Mistake." *East Asian Science, Technology and Society: An International Journal* 1 (1): 99–110. https://doi.org/10.1215/s12280-007-9004-7.

7 Intra-comparison: Translation, carriers, and mediators

Introduction

In the previous chapters we spent some time in Lombardy, Catalonia, and in the Brussels-Capital Region (BCR). We learned about the pilot activities of our TRANSFORM project partners working on participatory agenda setting and towards a push for deliberative democracy, in citizens science projects on the issues of waste collection and health, and on co-designing recommendations for regional development. We showed how all of these pilot projects are set in particular research and innovation (R&I) ecosystems while also trying to influence or even re-shape these ecosystems according to responsible research and innovation (RRI) principles. We saw how this is attempted through influencing both the processes as well as the content of regional innovation strategies, in particular their smart specialisation strategies (S3).

In this chapter we will stay with our colleagues from the different regions a bit longer before providing a broader European picture. We will revisit what we learned in the previous chapters but take a more comparative approach. What we saw so far is a great diversity of practices under the broader RRI umbrella. We opted for describing this diversity in terms of "translation," a concept that allows us to trace both the symbolic and material elements of such transformations, to focus on how knowledge is produced in particular political settings (Callon and Latour 1981; Callon 1986; Soneryd 2015; Soneryd and Amelung 2016).

We can now make full use of this concept by bringing the different translations together. Through this comparative perspective, we can also zoom in on the ways in which certain translations become stabilised. The premise here is that it is not at all arbitrary why certain versions of RRI are successful in a particular region while others are side-lined. The activities of the TRANSFORM project clusters take place within certain organisational structures with their own histories and power structures. They are designed, organised, and carried out by a set of actors with their own preferences when it comes to the methods to apply and the issues to tackle. These are powerful "organizational carriers" and different "normative and symbolic systems" (Soneryd and Amelung 2016)

DOI: 10.4324/9781003371229-7

that shape the translation of RRI in the regions. As we laid out in Chapter 2, such systems can be taken-for-granted beliefs and unquestioned truths. Other authors also point to the importance of institutional settings, zones of standardisation, and issue spaces in that regard (Chilvers, Pallett, and Hargreaves 2018). On a broader level, there are "systemic constitutional stabilities" such as legal frameworks, infrastructures, imaginaries, established social practices, and collective forms of public reason (Jasanoff 2005; Jasanoff and Kim 2015; Chilvers, Pallett, and Hargreaves 2018).

Looking for such carriers simply means acknowledging that RRI pilots don't start from scratch and do not take place in a vacuum (Kjølberg and Strand 2011). There have been initiatives before as there are most likely initiatives in parallel. Jason Chilvers and his colleagues in their work on the UK energy system transitions in a similar way direct attention to what they call "wider spaces of participation" and "constitutional stabilities" (Chilvers, Pallett, and Hargreaves 2018).

Thus, in comparing the different cluster activities we will pay particular attention to such organisational carriers, the wider spaces of participation as well as the constitutional stabilities within which they take shape and which they hope to re-shape. RRI, after all, at its core is about the attempt to re-think and re-shape cultures of research, innovation, and their governance (Owen, Macnaghten, and Stilgoe 2012; Stilgoe, Owen, and Macnaghten 2013).

As a part of this focus on institutional, organisational, and political environments of the TRANSFORM pilots and on the attempts at influencing them, this chapter will also touch on a set of activities that go beyond the core project work. These are activities that allow for the TRANSFORM activities to become "sticky" and have a legacy within the different regions. This means exploring work that has as its aim caring for, nurturing, and maintaining relationships that enable certain project activities to take place in the way they do. Within RRI discourses such ideas have been described with the notion of "care" (Kjølberg and Strand 2011). Stilgoe, Owen, and Macnaghten (2013) explicitly relate this notion to the governance of science and technology:

> Responsible innovation means taking care of the future through collective stewardship of science and innovation in the present.
>
> (Stilgoe, Owen, and Macnaghten 2013, 1570)

Responsibility here is framed in terms of *taking care of the future*. This is imagined to be done by what they call *collective stewardship* of science and innovation. Collective stewardship hints at a more democratic idea of R&I governance. One could make the argument that this is a call for making R&I as well as their governance "more" collective – read democratic – than they are in their current state. This focus on care can also be read as a critique of certain temporalities that are an – often implicit – part of innovation models. They way

Stilgoe and his co-authors discuss RRI as *taking care* has at least two very interesting temporal implications. To start with, they understand care as long-term engagement, which is rather different to short-term-oriented or one-off decision-making models of governance that follow a logic of choice. In addition, following to a logic of care in R&I practices (and their governance) becomes a question of timing. When does a certain activity or engagement with particular technoscientific issues, objects, or fields take place? What are windows of opportunity? At what point in the process of producing knowledge and developing innovations are different actors expected to become responsive to each other?

This can be read as a critique of more mainstream models of governance of technoscientific innovation insofar as they very often tend to intervene at the very end of these processes. Following a logic of care in contrast means early – or "upstream" – intervention and collaboration during the whole process. The objective of this mode of working is "to emphasize caring responsiveness in technoscience" (Puig de la Bellacasa 2011, 87).

Already this brief description of how *care* is discussed within a broader RRI discourse shows how this resonates with the work the different TRANSFORM clusters are committed to in their activities and also with the aim of TRANSFORM as one overarching SwafS-14 project. In their translations of RRI into territorial R&I ecosystems, the clusters develop different forms of care, meaning different ways of making territorial R&I a more collective endeavour.

This, so the argument goes, can contribute to moving beyond modes of governance aimed at prediction and control since these are considered to be part of the problem, that is, the multiple interconnected crises we are facing today (Guimarães Pereira 2015). In contrast, governance guided by principles of care focuses on "adaptation, ongoing tinkering, fine-tuning, and repair of processes and products by users situated in their settings" (Arora et al. 2020, 248). In accounts like this one, RRI and care are put in dialogue with the idea of maintenance, which has more recently entered debates about science, innovation, and the governance of R&I systems. In its everyday use *maintenance* refers to rather mundane practices like making sure that machines keep working. Maintenance, however, can also be thought of as an alternative mode of thinking about innovation. Framing innovation in terms of maintenance means focusing on caring for what is already there instead of fetishising the new (Vinsel and Russell 2020). Conceptualised like this maintenance can be seen as an alternative to more mainstream ideas of innovation (and its governance):

> In some ways, maintenance is the opposite of innovation. It is the practice of keeping daily life going, caring for the people and things that matter most to us, and ensuring that we preserve and sustain the inheritance of our collective pasts. It's the overlooked, undercompensated work that keeps our roads safe, our companies productive, and our lives happy and secure.
>
> (Vinsel and Russell 2020, 14)

Like the idea of *taking care of the future* maintenance also has a temporal dimension. Vinsel and Russell mention *inheritance* and call for *preserving* and *sustaining* that which is already there. While they mainly talk about technologies and innovation, we believe that these notions are also a very useful lens to think about regional RRI projects and thus the work going on in the different TRANSFORM clusters. This work is not only – maybe in some instances not even mainly – about the creation of something completely new. Very often the work in the clusters builds on and re-shuffles what is already there: Extending networks, nurturing relationships and in doing so slowly transforming cultures of responsibility in innovation governance. Importantly, this is work that often goes unnoticed, is invisible or pushed to the margins, especially in standard project evaluations.

Stressing the centrality of maintenance and care also enables us to write differently about the impact of the regional pilots and the TRANSFORM project overall. How is that so? Often, becoming responsive is translated as having *impacts* and *benefits* in projects funded by the EU SwafS programme.[1] This can be seen as a consequence of accountability measures put in place as part of European Union (EU) funding schemes for R&I. *Impacts and benefits*, unfortunately, are notoriously difficult to measure, especially when object of measurement is something as abstract as *transformative innovation governance*. One possible explanation for this is that RRI and governance for transformation more broadly is not well suited for top-down, command, and control intervention logics but is better understood through network approaches and self-governance (Strand and Spaapen 2021).

In this chapter we will take a comparative perspective on the three different clusters, look at the different translations of RRI we found, and describe how they are co-shaped with the various R&I ecosystems. Building on that comparison we will describe the work of the clusters and their ambitions of "impact" from a care and maintenance perspective. This also means pointing towards some of the tensions that this kind of work creates in the different regional clusters when it comes to ambitions of long-term impacts and benefits as well as with aspirations of transformation.

RRI goes territorial – a comparison of translations

In this chapter the aim is therefore to compare the different territorial translations of RRI. We will do so by pointing to the regions' answers to the question "What is *good* engagement?" The different answers to this question will point to differences as well as similarities with regard to the models of engagement and the rationales and (imagined) purposes of engagement, the different subject positions available, and in relation to ideas about acceptance and trust.

Following that, we ask how our colleagues talk about the legacy and impact of their activities and pilot projects. Through these accounts, we will discuss their theories of change – some implicit some very explicit – as well as how they conceive of potential innovation pathways.

The impact pathways and theories of change expressed by our colleagues will point us to the themes of *pilotification* and *maintenance work*. We will talk about piloting or showcasing of certain methods as a particular carrier of RRI in the different regions and also explore how a particular kind of work is necessary to make this carrier possible: Maintenance work.

Summing up this chapter we will compare the carriers and mediators of RRI in the different regions. In doing so we will come back to the distinction introduced above between methodological, organisational, institutional, and constitutional carriers and mediators.

What is good engagement for the different regions?

As we have shown in the previous chapters, the different clusters translate RRI in diverse ways. This concerns the methodologies they apply as well as the rationales and (imagined) purposes of this kind of work. This diversity also becomes nicely visible in how our colleagues talk about their ideas of what constitutes *good engagement*. They talk about real and fake engagement, about staying *on the ground,* and about the value of *being of service* as being core to their work. In what follows we would like to unpack these different ideas and see how they provide a first peak into the different regional translations of RRI.

As we did before, we would like to start in Lombardy. When we travelled to Milan to have some conversations with our colleagues, they were very keen on working with policy- and decision-makers in the field of R&I. Real engagement for them means working with "the people in charge of decision-making in governance and in the government, in the governments to govern R&I (Int_05).

The position of Fondazione Giannino Bassetti (FGB) within the Lombardian innovation ecosystem gives them access to actors within the Lombardy Region who are willing to participate in the pilots. This framing of RRI is thus closely tied to how the network is set up and to the long tradition of FGB working with the Region on issues of responsibility. This collaboration is also enshrined in the often-mentioned "Legge Regionale 23 novembre 2016, n. 29," which states that to strengthen regional innovation and the competitiveness of the system, a culture of RRI needs to be established through the dissemination of and experimentation with innovative methods and processes. As we already mentioned in the chapter on the Lombardian cluster, this law also provides the legal basis for the so-called "Forum for Research and Innovation," for which FGB is formally recognised as a supporting body. There is thus what Chilvers, Pallett, and Hargreaves refer to as "powerful systemic constitutional stabilities" (Chilvers, Pallett, and Hargreaves 2018) to the work FGB is doing in Lombardy.

Working with actors in key decision-making positions is thus one key criterion for good engagement. But there is more to this: Good engagement also needs to be *real* in the accounts of our Lombardian colleagues.

This focus on working with policy- and decision-makers in order to achieve *real* engagement in the form of some actual consequences of the work, however, also means that the citizens and the subject positions available to them are different compared to the work in the other clusters. In the participatory agenda-setting process, available subject positions appear mainly as carriers of region-specific needs. The aim of addressing them via survey and focus-group methods is to give the Lombardy Region more precise information about the needs of the different regions within Lombardy in order to improve their policies. Similarly, also in the citizens' jury, the participants are conceived of as carriers of a lived experience in the region and therefore as experts with regard to what needs to be improved. They are framed as "to ensure a diversity of voices, experiences and points of view"[2]. As such they are distinguished from the more technical experts who participated in the event to inform the citizens about issues like Big Data and AI, open data, and "some technological features of smart mobility."[3] The citizens then developed recommendations that were presented to Lombardy Region who "committed to taking citizens' ideas and suggestions into account in its future actions in the area of smart mobility and, should it not be possible to incorporate them (...), to explain the reasons for this."[4] While this certainly is state-of-the-art when it comes to citizens' juries, it is a very particular way of engaging citizens which is closely tied to the ambition to provide something *real* and to showcase a methodology to regional administrative actors.

The inverse of real engagement in the way our colleagues from Lombardy talk about their activities is *fake* engagement. Fake engagement is mainly characterised by not having any consequences whatsoever. And this is the second main principle of doing this kind of work in Lombardy: Producing something that has a consequence or impact on the R&I policies in the region. This impact, in the view of FGB, can be in terms of policy but also with regard to the very governance structures within the Lombardy Region as a long-term ambition is to find ways of establishing a more permanent citizens' assembly in the region.

In a similar way, our colleagues in Catalonia stress the importance of doing something *real*. However, their understanding of what is real and what they would demarcate it from is quite different from what we just described for the Lombardy cluster. While in Lombardy real engagement is very much focused on the outcomes and thus the consequentiality of engagement activities, the starting point in the Catalan cluster is the challenges they are working on.

As we see in quotes like this one, the focus is on the challenge. Both the endometriosis and the waste collection cases emerged out of the Catalan think tanks, which meant that they presented problems that people where really struggling with at the time. And this was exactly the idea that guided the work of this regional think tank: To find a challenge that really matters to the lives of people on site. This in turn is what promises some form of impact or (hopefully positive) outcome in the end. Also, our colleagues stress that involving people in citizen science approaches will allow for being "much more innovative" (Int_01).

In the Catalan cluster, being attentive to the particular context in which the pilots take place and making room for the lived experiences of the citizens was a core tenet of what it meant work on something real. For example, the waste collection pilot showed that there is not a single solution that is suitable for every neighbourhood. Therefore, the aim was "to co-create with people like the ideal waste collection system for their neighbourhood. (…) So also people in charge needs to be flexible and understand that maybe not all one solution fits all" (Int_01). *Co-creation* is here presented as the other of simply *implementing* a solution without involvement of the people affected – the residents in the case of the waste collection system and the patients in the case of endometriosis.

The citizens thus are conceptualised not merely as experts on certain needs and carriers of opinions, but epistemic actors who actually can contribute with regard to the specificities of the challenges – its situatedness if you will – and also when it comes to the best ways of addressing them.

This position should not be confused with a naïve romanticisation in the sense of *the citizens always know best*. It merely means that proper engagement needs to start with *real* problems that are affecting people and that in order to solve these problems these same people need to be involved. As one of the cluster members put it: "Citizen science is not the objective. It's a means to something. It's to do something" (Int_02).

This is less about governing technoscience and more about governing the ways in which technoscience is interwoven with the lived realities of the people in the territories.

However, the thinking and acting of our colleagues in the Catalan cluster do not stop there. There is also governance and technoscience in the more "traditional" sense of the terms. Firstly, similar to the activities in Lombardy, the work this cluster is doing also includes translation in the sense of Callon and Latour, that is, the enrolment of allies into a network of people who already know what good engagement is. It is

> an opportunity also to start talking to each other inside the, the municipality (…) and to change a little bit the waste, the people that work in waste management.
>
> (Int_02)

In quotes like this one our colleagues would talk about different administrative entities including the municipality but also Hospital de Sant Pau or the Generalitat. So, good engagement is about community building both within and beyond the administrative partners. The overarching aim is to have engagement to make better policies:

> I want to use citizen science to open the views of policymakers to understand the perspectives of the citizens and to make better policies and then it's a different question.
>
> (Int_03)

This is how good engagement is connected to actual policymaking and governance of technoscience. Citizen science here is framed as a means to introduce contingency in the sense of getting to know and *understand* different *perspectives*. The issue then becomes less one of crowdsourcing, which is often prominent in citizen science approaches, but one of framing the issues to be solved. This way of framing problems differently introduces contingency which is supposed, in the view of the members of the Lombardy cluster, to appreciate the need to "work in different ways, public administrations with universities, with companies, with citizens to address the challenges in different way" (Int_03).

Core to this conceptualisation then is that *good* engagement – and in extension *good* governance of technoscience in the Catalan translation of RRI – appears as a willingness to *transform*. Good engagement needs to aim for actual transformations:

> I connect to people who want to be transformative and then we are a very growing network because there were more people in different fields working on that.
>
> (Int_03)

In the BCR cluster we observe several distinct yet related translations of RRI. Generally speaking, RRI is framed as some version of citizen participation or deliberative democracy. In the BCR cluster, this takes the shape of policy co-design (in the Unsold Food pilot) and co-design of (technological) innovations of student-innovators from UC Louvain (AquaSens and Algorella). In addition to these pilot activities, the BCR cluster worked towards the development of the so-called "Spheres-protocol," an overarching activity intended to analyse and support projects with regard to RRI.

These translations are also connected to different answers to the question "what constitutes *good* engagement?" As we saw in Chapter 6, good engagement in the BCR cluster centres around an ethos of *being of service*. Being of service here first and foremost means working with bottom-up initiatives and provide support with regard to emerging conflicts or in terms of support from public authorities. Crucially, engaging with initiatives that work on the issue of unsold food is intended to create conditions in which locally situated knowledge and lived experiences of food production and consumption systems can enter policy- and decision-making spheres. This idea of being of service thus carries connotations of community development but manages to weave in the administrative partner INNOVIRIS via their need to re-think evaluation criteria for the selection of regional innovation projects and to find ways of making such initiatives more "sustainable." Be Participation – and this is a crucial difference to the Lombardy cluster – form these kinds of engagements exactly because it is not part of the extended governance structure yet still entertains connections to the policy realm.

We saw an additional version of *being of service* that is enacted in the work of UC Louvain as well as in the overarching Spheres protocol. Here, the service is mainly aimed at researchers and intended to help them anticipate issues that they would not have otherwise recognised, as exemplified in statements like this one:

> And what is interesting is basically what we tried to also explain to the researcher is that you can develop a very nice innovation that will measure the quality of water but if you don't pay attention to the perception that citizen have, you, they will maybe they will say okay, fine, I can test the quality but I will still use the bottled water.
>
> (Int_10)

The work of this cluster thus followed a model of engagement that was in a sense very much led by the citizens. They had already organised into initiatives around the issue of unsold food, bringing together themes of poverty and socio-economic inequality with questions of environmental protection and use of resources. Be Participation's role was mainly one of providing the knowledge and experience with regard to co-design and multi-stakeholder engagement methods. The participants in the unsold food pilot thus were "used" to showcase the added value of these methods to the public authority in the form of INNOVIRIS. At the same time, however, the engagement was still used to show some potential paths towards solving the issues and challenges. Our colleagues from Lombardy, however, were keen to point out that their mandate was not one of providing or elaborating solutions as they were only organising piloting activities. In a similar manner also the work of UC Louvain was intended to engage multiple stakeholders in the development of PhD students' innovations while at the same time aiming to convince university managers to adapt the university's doctoral education towards a more holistic view of innovation.

The actors that participate in these co-design workshops are framed by our colleagues as *disadvantaged publics*. Disadvantaged as it is used here means that these actors are not part of a powerful elite that is able to shape policies and decisions. As such, they occupy different subject positions than participants in the other clusters, as they are the ones framing the problem. They have agency in the sense that they set up an initiative independently from public authorities and already go into the engagement with their own issues and challenges. In addition, these are people who wouldn't usually attend top-down focus groups.

The PhD students from UC Louvain in contrast occupy a different subject position. They are framed as innovators in need of input from citizens with regard to the usability of their innovations as well as to potential ethical and social implications. As such they are sometimes also perceived as being in need of protection from the citizens, especially if they should be granted the mandate to cancel certain projects if they are seen to be dangerous.

Interestingly, in these conversations also the notion of *acceptance* came up on several occasions. And indeed, *acceptability* is a crucial part of conceptualisations of RRI, as visible, for example, in this quote from von Schomberg (2012):

> Responsible Research and Innovation is a transparent, interactive process by which societal actors and innovators become mutually responsive to each other with a view to the (ethical) acceptability, sustainability and societal desirability of the innovation process and its marketable products (in order to allow a proper embedding of scientific and technological advances in our society)

There are at least two ways in which one could think about acceptance in relation to RRI work: On the one hand there is the more educationalist and awareness-raising type that aims for increasing the chances that citizens will not reject a technology or policy decision due to being informed about it. This is of the type "if only they knew more about X and understood it better, they could more easily accept X." On the other hand, one can think about acceptance or acceptability as creating processes that allow citizens to exert some actual influence on innovation pathways or certain decisions. This then would mean providing conditions under which R&I processes and their outcomes become in fact more *acceptable*. This latter version of course is closer to the core idea of RRI.

In the conversations we had about the cluster activities we addressed these nuances of the notion.

> I mean for the moment in Brussels you have a campaign about air pollution but what will be the, the perception of the people that have been employed in the in the campaign and the people that have not been employed in the campaign. Is there a difference. If there is a decision for example to say okay, we will limit that thing to increase the, the to increase the quality of air, what would be the acceptance in the two different groups.
>
> (Int_11)

While this is close to the traditional idea of the problem of acceptance, the term also can be understood as the other of *social contestability*. Framed like this, Spheres turns into a *vehicle* for working with small and medium-sized enterprises (SMEs) on social contestability issues. Similar to the challenges for the PhD student-innovators, the public here is framed as an obstacle for innovation or as an entity that will contest what innovators or researchers are doing.

> So they can focus on the technological aspect but they, they lack other competences that they could need in the lifecycle of, of the innovation at one moment. Sometimes very early, sometimes a bit later. And strangely, they are working like they hope that succeeding on the technological part will solve each and every problem. And when we were asking them yes but what about social contestability because you will end up by doing that but it's not sure that your neighbour because some firm have just 50 metres are just at, sorry. Yes. 50 metres of their neighbours, so you have that not in my backyard problem and so on. What about your future social contestability They say well what is contestability. How do we manage that. And actually in the lifecycle of an innovation, it appears one, the problem appears. So they discover that you have Greenpeace that is knocking at your door and saying you have to stop all the activity here because there is an environmental problem. And just replying "oh, we are a firm recycling waste" that, that is not solving the problem. So we discovered that we had to have a vehicle to a kind of what we called at that time one-stop shopping where they could find some competency that they don't have inside, OK and that they could use that very early in the innovation process, OK.
>
> (Int_12)

Here we see again a service logic attached to RRI work. However, this is slightly different from the one we talked about before. The objective of bottom-up engagement here means helping the innovator "to anticipate some questions about their impact on the environment, on the public health, equity" (Int_12).

After this comparative summary of the cluster's ideas about what constitutes *good engagement*, we will now turn to the stories and practices of creating impact through the TRANSFORM RRI initiatives in the different regions.

The remains of a pilot – thoughts on legacy and impact

As we just saw, the TRANSFORM regional clusters differ in what constitutes good engagement as a part of RRI activities for them. But this is not where the differences stop. There is also some diversity in how they think about their own legacy. The starting point, however, is the same. The project's Description of the Action (DoA) introduces the following causal narrative for how impact is expected to be generated.

The impact pathway(s) to be studied by WP7 can be summarised in the following simplified form:

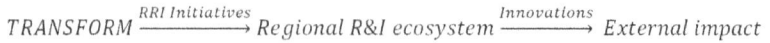

$$TRANSFORM \xrightarrow{\text{RRI Initiatives}} Regional\ R\&I\ ecosystem \xrightarrow{\text{Innovations}} External\ impact$$

Figure 7.1 TRANSFORM causal narrative.

The basic idea expressed in Figure 7.1 is that TRANSFORM organises and conducts RRI initiatives with and for the regional R&I ecosystems. These initiatives will then impact how innovations get produced and disseminated, which in turn leads to external impact. Arguably, this is a very simple narrative that will most certainly not be able to capture the actual impact pathways of the project.

From this starting point, therefore, it was necessary to complicate things by adding more elements and introduce different relationships. This made the picture more nuanced.

In the project's own monitoring and evaluation guide, we introduced this graph (Figure 7.2) that depicts the causal narrative of TRANSFORM project activities. It shows how a set of actions (TRANSFORM actions) – these are RRI initiatives and tasks within the project performed by actors working for the project – will result in *project outputs*. From there, the TRANSFORM actions are expected to create effects in the regional clusters. For instance, one can imagine that they will influence innovation pilots or innovation policy-making processes (both in terms of their content but also regarding their very existence). One might also assume, or hope, that these actions have effects on the actors in the regional clusters. They could, for example, improve the RRI competence and/or awareness and also shape networks of different actors. This would be the project outcomes in proper project evaluation terminology.

In the next step, the actors in the regional clusters perform actions such as planning, organising, and conducting RRI initiatives in their regional R&I ecosystem. Examples are revisions of regional innovation policies, the inclusion of RRI elements in RIS3 plans, and also processes that may or may not fall under point 2 above, such as innovation pilots, competence building, etc. In project evaluation terminology, the results of these actions belong to the class of project impacts.

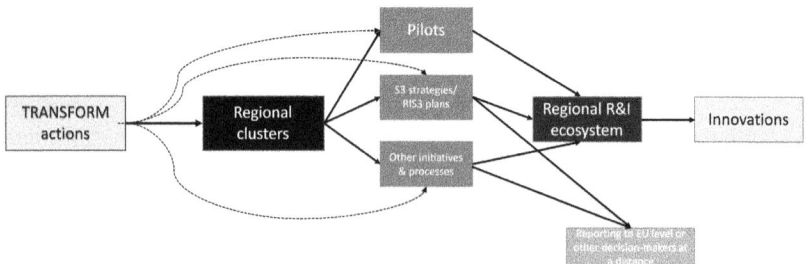

Figure 7.2 The project's theory of change as sketched in Deliverable 7.1.

Starting from this – admittedly crude – theory of change, we talked to our colleagues about their view of the relation of project activities, outcomes, legacy, and impact. In this section, we want to contrast what we learned in the conversations we had with the Brussels team with insights from our trips to Milan and Barcelona.

Lombardy – feasibility and actionability

The TRANSFORM Lombardy cluster, in close collaboration with the regional administration represented by Lombardy Region and Finlombarda, designed and conducted a participatory agenda-setting process and a citizens' jury on the theme of regional smart mobility. As we saw in Chapter 4, our colleagues followed a list of ambitious aims described in the grant agreement, which envisioned these activities having an impact on Lombardy's R&I policy. Recall the TRANSFORM Lombardy cluster's goals from Chapter 4:

Set-up and carry-out a multi-stakeholder engagement process through participatory research agenda setting approach.

Include concrete suggestions, visions and opinions from citizens and local stakeholders in the next Lombardy Region Three Years R&I Strategic Plan, aligned with regional S3.

Develop a detailed operation plan which describes the whole process and can ensure the replicability of the approach in Lombardy and beyond.

Foster novel and transparent governance relations within the regional R&I agenda setting.

As these aims already make clear, the goals are not only about influencing the shape of regional innovation policy but point to something more fundamental. *Fostering novel and transparent governance relations* indicates an ambition to re-shape regional innovation cultures. In one of the final outputs of the project, the so-called "Strategic roadmap for the implementation and support of territorial RRI through Participatory Research Agenda Setting,"[5] our colleagues from Lombardy consider these aims achieved at least to some extent. Recommendations from the citizens' jury were *delivered* to the regional administration for further consideration. This is of course a notable outcome, but would not constitute a ground-breaking achievement in itself. In addition, however, governance was put on the agenda:

Thanks to TRANSFORM, regional S3 (2021-2027) has set participatory governance as one of the main challenges for the region. The strategy explicitly mentions TRANSFORM experience and dedicates a full chapter to RRI principles.

(Lombardy Strategic Roadmap)

Participatory governance is now recognised as a *governance challenge*. This might be read as a rather generic statement. However, RRI entering regional R&I policy discourses does not come out of the blue but rests on ongoing maintenance-work and a certain understanding of impact pathways.

In conversation our Lombardian colleagues expressed a basic idea of impact pathways premised on producing meaningful outcomes to the citizens by different partners from industry and the Region. Such and idea might sound trivial, but there are interesting twists in how this understanding of impact pathways relates to a particular translation of RRI in the regional cluster. Here is one instance in which our colleagues talk about their activities and what they achieve:

> The technology roadmap is addressed mainly to industrial players. And we think that because of the topic we can enlarge the team on other elements connected to artificial intelligence that are more relevant for citizens and that can, that Lombardy Region can use for farther policies that are in the pipeline as well to regulate and govern artificial intelligence in the activities of the regions or in funding, opening call for proposal and project connected to artificial intelligence, taking into account the elements stemming from the results of our participatory activities.
>
> (Int_05)

What is expressed in quotes like this one is the idea that deliberative democracy gets a footing in regional governance through *proof of concept* like practices. This links back to the approach of *piloting* we described above and goes beyond raising awareness as the aim is to convince ever more actors within the Lombardy Region and to give them the skills to organise such processes themselves.

> I hope that this experience will be the first experience of a long journey about investing in terms of resources, that means personnel but also money, fund to ground other sound citizen engagement processes. (...) I think was the sort of proof that something nice can be done.
>
> (Int_05)

Having a good proof in this way then is imagined to have consequences within regional administrations: Firstly, there is an element of *getting used to* certain ways of working. This resonates with insights form social science about the importance of routinised ways of working (Schatzki, Cetina, and Savigny 2005; Shove 2010) but also with insight from transformation studies about certain "institutionalized habits of thought" (Jasanoff 2003) that shape how we see and act in certain environments. These elements are also expressed when our colleagues stress the importance of their administrative partners getting to doing something, because then "they do it because they are used to do it and because they need to do it" (Int_06).

Interestingly, this goes beyond a simple understanding of cognitive deficit and the need for awareness raising that follows from it. Instead, it points to an element of capacity building or developing skills that involve more institutionalised modes of collective action. Even if their partners understand how to incorporate elements of deliberative democracy in the development of innovation policies, and into questions of resource distribution, this does not necessarily mean that the same elements are also integrated in how the R&I ecosystem operates. Piloting and proofing, in order to have an impact, in this understanding strongly relies on doing things together, as explained by the administrative partners of the Lombardy cluster:

> So, at the moment in, in we are not ready to do it alone. We need a little bit more support also after TRANSFORM and maybe it will be possible to do it with JRC or I don't know. Maybe, obviously we will go on working with [NN] and with Fondazione Bassetti absolutely. Because they are for us a very important reference here in Milan
>
> (Int_08)

Statements like this one point to a step-by-step change of governance cultures. They need to actually learn to do things themselves. Not only for the purpose of conducting these activities but also to better understand how and in which instances they are best applied. The regional administration needs to learn because only they know what to do when.

This is where the focus on *real* engagement comes from. Only if there is something that has real consequences – independently how big they might be – will the Region, representatives from the industry, and citizens be convinced that deliberative democracy is not just a "waste of time, public money" (Int_05). What is at stake here is not only the success of the project, but that of the Bassetti Foundation overall and as such the chance to infuse the Lombardian R&I ecosystem with more responsible innovation practices and structures:

> Or at least at the end the face in front of citizens working and talking with them in the workshop, it's me. It's not Lombardy Region. It's me. It's Bassetti Foundation. We have long history for responsible innovation. I don't want to have fake participatory process so just to say or to have a deliverable or to tweet on Twitter, 'oh we have made this wonderful workshop collecting recommendation that no one in the world will read and use somehow.'
>
> (Int_05)

What is needed to avoid this fate is a careful balance of the needs of the different actors. There need to be consequences of the participatory activities, however, they cannot be such that the Region and industry representatives would be de-incentivised from relying on such approaches in the future. This in turn means that there will be mostly "incremental change" (Int_07) as one member from the Lombardy cluster told us. In that way, as one colleague put it, even if TRANSFORM is merely "a little project [...] it can be effective in providing a policy a genuine policy change in the way the regional public administration work" (Int_07). This understanding of impact thus goes beyond shaping the *what* of particular innovation strategies. It is about re-shaping practices and cultures.

Finding this important balance influences how such processes can be organised, but it also makes some areas more likely objects of deliberation than others. In innovation areas that are well-funded and where there is little competition for scarce resources, for example, are more likely to be aligned with citizens' needs and preferences.

In that way the particular Lombardian translation of RRI is directly connected to views about impact pathways:

> Feasibility and actionability are definitely my key words for the process in, in Lombardy.
>
> (Int_05)

Catalonia – projects don't change policies

The overarching objective of the Catalan cluster in the TRANSFORM project was to support a "transition to a more competitive, sustainable and inclusive economic model."[6] The work in this cluster relied on citizen science as the approach for integrating RRI principles and methodologies into the region's RIS3 plan, the RIS3CAT 2021–2027.

This meant experimenting with citizen science approaches involving actors from the quadruple helix, meaning from public administration and academia, but also corporate actors and civil society members. Importantly, this meant involving these actors from the very beginning in designing the pilot activities.

In the grant agreement this translates into the aim of integrating "participatory strategies and citizen science methodologies in the new RIS3CAT strategy," to re-shape "ongoing RIS3CAT funded projects from the triple helix towards the quadruple helix," and to achieve that to design and conduct "multi-stakeholder engagement to embed public participation and citizen science in ongoing RIS3CAT projects."

Interestingly, and here we enter even deeper into the realm of theories of change, the official aims of this cluster's work include using the project results as "new selection criteria for future calls for new Communities RIS3CAT" as well as for the "future specialisation strategy of the Region." Through these activities, the cluster expects to build "novel and transparent governance relations within the regional R&I agenda setting."

Already in these aims we can see several different ideas about impact pathways. Obviously, since this is the overall objective of TRANSFORM, there is the expectation that there will be an impact on the RIS3CAT through interaction between the research and administrative partners. However, things are more nuanced already on this level. We can see the idea that innovation policy, and importantly, processes and mechanisms by which innovation is governed are impacted by showcasing novel approaches.

However, "[p]rojects don't change policies" (Int_03) as one of the administrative partners explains. This shows an interesting tension with the project's promises and the expectations created by European funding structures. The argument here is that "it's important because sometimes we misplace the things. And the people saying that projects will change the whole policies are lying a little bit because usually it's not the case" (Int_03).

What this means is that there is an awareness that TRANSFORM as a project is situated within a vast ecosystem of ongoing projects and initiatives as well as historically in a lineage of certain activities.

> Yeah, because for me Transform is part of the process. It's not starting and finishing. It's I don't know. No because the idea is that with Sant Pau with the hospital that they use the results to change the protocols that the Agency of Quality of Catalonia [Agency of Health Quality and Assessment of Catalonia; T.V.] is rethinking how they can use citizen science to improve health policies. And maybe we are going to do another project also with [NN] so I don't see, I don't want to think about the projects and after the projects because what I am doing is not for the project what I would do.
>
> (Int_03)

The way our colleagues talk about the project, it is always already a *predecessor to be* when follow-up projects are discussed and imagined. One colleague from the public administration states that: "it never works if you build things from nowhere. You need to have, and this is because these projects have no money for anything, so you need to connect with something that is already happening in the territory" (Int_03). How then does this translate into the cluster's ideas about the impact of their work?

One of the more prominent impact pathways in the accounts of our Catalan colleagues concerns their regional think tank. Initially, this body was rather large and brought together people from academia, public administration, the business sector, and civil society. The idea then was to initiate a cultural shift in how people think about and govern innovation by means of *demonstration.* This is an idea we also see in the other clusters and which is a common way of thinking about the impact of piloting. There is a focus on *replicability* in these accounts that is very interesting as it provides an additional element to common discourses about added value that is similar to the framing of the Lombardy cluster colleagues. People need to see a clear path for how they can apply what is presented. Showing added value then might be necessary but not sufficient to achieve an impact.

In addition to this idea of impact by demonstration, there is also a notion of creating new links and thereby creating new dynamics within the regional innovation ecosystem. This is then about "change in the people participating" and an ambition to support "different dynamics in the networks and collaboration for example" (Int_03).

The aim for shifts in practices and also in innovation cultures is clearly expressed by our colleagues when they state, for example, that they want to push their partners to *think differently*:

> And that's trying to push them and push them a little bit to the municipalities or even the doctors of, of Sant Pau to think differently, no.
>
> (Int_04)

An important condition for that, again similar to Lombardy, is to create trust within the public administration. This, in the viewpoint of the Catalan cluster, is achieved by showcasing very specifically what can be gained by using certain methods, "(n)ot just the data but also on the dialogue that they generate (...)" (Int_02). It is not only ways of thinking, however, through which our colleagues in Catalonia aim at shifting cultures within public administration. It is also by creating new links between departments which "are really happy (...) because they don't work together usually" (Int_02).

Citizen science works here as an *umbrella-term* that allows the cluster members to introduce novel things and add to the tools they already use. In that way, changes within the administration can be initiated:

> Umbrella, yeah, where citizen science can be incorporated or how the public administration, the regional public administration can foster within the tools that they have in, in that sense. They foster this kind of processes. The institutionalisation is important for me.
>
> (Int_02)

Building on this work, plans or ideas for future collaborations are already being made. One potential avenue for developing projects that our colleagues talk about concerns the metrics that are used by the public administration to evaluate certain public sector systems.

> We started reflecting with Aigües what is the process of the methodology used to develop these PREMs for example and it was, we arrived to the

idea that this is something that we cannot do within Transform. It's too big to develop that but it's a very interesting thing because it's a tool that public administration that evaluates the health system for example can have a metric on that.

(Int_02)

In addition to creating impact through the think tank and via the work with the administrative partners, collaborations with universities are described as a way to create impact within the regional R&I ecosystem. This resonates with work done in Brussels by UC Louvain, who also focuses on higher education as a sector through which to shape innovation cultures.

A final element in the theories of change expressed by the actors in of the Catalan cluster concerns standardisation:

We are also working on the standardisation of the methodology since 2019 now and the standard is quite advanced so it could be like the first technical standard with the name citizen science on, on it. So, this will also help I think to the whole community to get recognition and credibility and validation of data and so on. So, we are on that and once the standard is over, hopefully then the authorities will update the methodology as official and maybe we will inform some new regulations also.

(Int_01)

This is a clear ambition to re-shape what Chilvers, Pallett, and Hargreaves (2018) refer to as the "wider spaces of participation," in this case ab attempt of creating zones of standardisation. The hope is even to push for new regulations and thus to establish RRI or citizen science on a constitutional level. This is clearly very ambitious but nonetheless a promising strategy for creating impact as can be seen also in the Lombardy cluster and how they use the regional legislation to stabilise RRI in regional innovation governance.

What we see in Catalonia is a nuanced picture of impact that brings together a diversity of elements in a multi-faceted theory of change. Also, there is a clear ambition to change the way in which actors in the regional R&I ecosystem work and think about innovation and about solving contemporary challenges. This is combined with a keen awareness that one single project will not do the job. Therefore, importantly, when the partners in this cluster talk about legacy or impact, they rarely use TRANSFORM – or any single project for that matter – as a reference point. They rather turn mainstream impact narratives on their head: They start from what they want to achieve and then talk about the project as one small piece in their plans and ambitions.

Brussels-Capital Region – being of service in a complex R&I ecosystem

The BCR cluster describes several aims in the grant agreement that focus on establishing particular processes and procedures as well as on a set of distinct themes. As mentioned in the previous chapter, these aims include the following:

- Set up and carry out a multi-stakeholder engagement process through design thinking approach.
- Address concrete community needs, emerging from a bottom-up participatory process involving at least 200 citizens and carried out in two distinct Brussels districts.
- Engage citizens, local CSOs, and universities/SMEs in the co-creation of social innovation solutions in the field of circular economy, leveraging on the urban metabolism of Brussels: Identifying specific resources to be reinjected into the local economy rather than exported as waste, supply chains and local economic actors to implement solutions, and market demand to benefit and scale up the proposed solutions.
- Foster novel and transparent governance relations within the regional R&I agenda setting.

Central elements in these aims are multi-stakeholder engagements and bottom-up participatory processes. Initially, design thinking was to play a major role in the achievement of these aims. This approach turned out to be less influential than anticipated throughout the project – also due to some organisational changes in the cluster set-up. Like in the other clusters, in addition to engagement activities also changes within governance practices and processes were envisioned.

As we described in previous chapters, our colleagues in the BCR cluster were confronted with a highly dynamic and fragmented R&I ecosystem. This has led to several shifts and changes during the projects. The members of this cluster organised different activities and also made plans for a number of initiatives that did not fully come to fruition. Nonetheless, also the plans that were abandoned and replaced during the project illustrate how impact pathways are imagined in the work of this cluster.

The account of impact pathways that is closest to what is stated in the project description, and to how we depicted causal narratives in our work, centres around the relationship between Be Participation, UC Louvain, and INNOVIRIS. One of the main ideas in the beginning of the project was to collaborate with INNOVIRIS to mainstream RRI-inspired practices beyond their Co-Create program.

But the idea is okay, they are already doing it in one specific program called Co-Create but the whole idea is to mainstream that and to have an impact on a more like broader level like in the, the way the regional innovation plan is built, in the way they evaluate the projects et cetera, et cetera. And not only in one specific stream of funding like, so. Yeah.

(Int_10)

This then is straightforward: The TRANSFORM project activities showcase the added value of RRI approaches to the formulation of R&I strategies and because INNOVIRIS as representative of the regional administration is part of the project, they can initiate shifts in their processes.

However, with regard to the TRANSFORM project, INNOVIRIS was rather reluctant to go beyond taking on board the pilot projects of the cluster in the form of recommendations, so our colleagues had to come up with alternative ideas of working with their administrative partners.

One idea focused on what our colleagues in Brussels referred to as "the grid." This grid basically consists of a set of evaluation criteria that are used by INNOVIRIS in its processes of assessing the quality of project proposals. These criteria are different for the different funding schemes that exist at INNOVIRIS.

This grid you're talking about it's something else. We made a call to see if there is some project that could be interested in to be advised by TRANSFORM, but with not a lot of success. And in that case, I used a grid we used for the evaluation of projects and including that includes the social and environmental aspects. (…) So, you have different kind of, how can I say it, criteria and that's a way we use the grid.

(Int_11)

During the discussions around the Spheres protocol that we described in the previous chapter, the BCR cluster thought about using this grid and then to try re-shaping them according to RRI principles.

Unfortunately, as the Spheres pilots didn't turn out as our colleagues had hoped they would, also the plan on working on the INNOVIRIS evaluation grid never fully came to fruition.

Throughout the lifetime of the project, therefore, the ambitions changed significantly as it was not possible to initiate any changes in the way INNOVIRIS is organising its funding schemes. Together with these ambitions also the models of impact shifted. The attention thus was directed to the issue of unsold food. This allowed our colleagues to organise what they refer to as "meaningful engagement." *Meaningful* is used here in the sense of providing a service to an ongoing initiative that is rooted in concrete problems of citizens in a certain area of the city. Responsibility here is translated as being *responsible for* something or somebody as in caring for them. In this case, this meant caring for local communities.

The main challenge that was addressed in the work of this part of the cluster activities was that some actors were interested in food-waste because they wanted to organise food for people in need, while others tended to also see this situation as a business opportunity. As it turned out, this project was also interesting for INNOVIRIS as it related to their funding practices. By funding some actors involved in the business of unsold food, they unintentionally created tensions between different initiatives and businesses working on the issue. This TRANSFORM activity therefore still has the potential to influence how funding decisions at INNOVIRIS are made in the future.

The ambition here – as our colleagues were keen to point out – was not to actually *solve* the conflicts between the different actors engaged in this issue. The aim was rather to bring the different actors at the table and show how a process of solving their issues could look like. The engagements indeed ended with a "sharing workshop" in which actors involved in the issue discussed how next steps could look like. This process involved citizens, non-profit organisations, and representatives of regional authorities and funding bodies.

The legacy of this work will most likely not show up in any future innovation plan in the sense of certain topics being highlighted or innovations that answer to precise citizen-needs being proposed. This model of impact instead aims at the practices and procedures of regional administrations and – in this particular case –funding agencies. It is about how projects are selected for funding and about the question of how to take into account what is already going on with regard to a certain issue.

In addition to the activities we described above, there was another strand of work within this cluster that was premised on a different model of how to create impact. This work that was led mainly by UC Louvain presents a quite clear albeit different idea about the legacy of its activities. To start with, the idea of our colleagues in Leuven is explicitly contrasted with a model of change through policy documents, something that is called *nice words* or *wishful thinking* and has been addressed in the literature as "buzzwords" by, for example, Bernadette Bensaude-Vincent (2014).

> We want to go out of wishful thinking about the beauty of citizen engagement, ok. We cannot afford that. And by the way, we had a discussion, or we made a presentation with [NN] at all the, how much, 37 responsible research projects in the research administration and they are convinced that it can be a vehicle for all the research at universities. So, each research project could be tested that way but then we have to be very efficient.
>
> (Int_12)

The theory of change presented here focuses on working directly with innovators on the level of a PhD education. The Spheres protocols we discussed in a previous chapter play a crucial role in this model. They are introduced into a PhD program as sort of *test* if an innovation can be what is referred to as *360 degrees innovation*:

> So, you address the work of the researcher, the innovation with citizens, with experts in public policies with academics. Also, academics or industry. So, we try to make it 360 degrees to understand what could be the reception of the citizens on the innovation, on the thing.
>
> (Int_11)

Working with PhD students using the Spheres protocols is imagined to convince the university management to re-arrange their PhD programmes around a different, broader notion of innovation. A notion of innovation that takes on board RRI approaches and principles but also takes on the legacy of technology assessment and ELSA/ELSI approaches by focusing on different sets of *implications* of technoscientific innovation:

> So there were for example students that engineer in at the engineering faculty that they have their first chapter which was about the technological aspects of material that they were developing for a building material. Then the second chapter was about the environmental implications and then they wanted the last chapter to actually consider the, the social implications of their theses. So there is already something that is that is being cooked sort of saying at the university and where Transform could actually have a, there is room for, for Transform to have a legacy in.
>
> (Int_12)

The impact of this would then be initiating a cultural shift towards RRI with the next generations of innovators-in-training in a sense. Proof of that in turn – in the view of our colleagues from UC Louvain – could be given to INNOVIRIS.

> Just something very briefly but we, we know that part of the or the Transform aim is to have an impact on, on policy and at the beginning it was great to have Innoviris as, as the main agency promoting innovation because the impact can be there. (…) I think that we can have some proof afterwards that can be given to them about the potential of including citizens in the innovation process.
>
> (Int_12)

Again, this is rather different from influencing the next S3 strategy with regard to particular themes and innovation areas or even in the way the S3 strategies are produced. Much rather, the understanding of impact visible in the accounts of our colleagues from UC Louvain focuses on long-term cultural shifts. Shifts that, in this case, will only impact the R&I ecosystem indirectly via the education of the next generation of engineers and innovators. Instead of innovation governance, we tend to see attempts at governing innovation governance (we will take up that thought in the next chapter.

Over the last couple of pages, we have talked about different understandings of impact and impact pathways. One important thing we learned is that the different TRANSFORM clusters all aimed at cultural changes within regional administrations as part of the broader territorial R&I ecosystems. Now, clearly this is an ambitious aim if there ever was one. Achieving such shifts in habitual practices and cultures takes time and a lot of showcasing and convincing. It also takes a particular kind of *work* that is not usually captured by project evaluations. Following scholars from the field of science and technology studies, we will refer to this kind of work as *maintenance work*.

Piloting and showcasing as maintenance[7]

In the previous sections, we told three stories about the activities in the different regions. Already there it became clear that our colleagues are doing something beyond the activities described in the official work package descriptions. As one colleague from Catalonia described what is going in the cluster, he ended with "there is more than that." Another member of the project consortium described their pilot project in a similar way as "technical but not so technical." Statements like that point to work that is done beyond what can be captured in the description of work in grant agreements and even beyond what can be measured in project assessments. In this section, we want to dive deeper into this *more than* and talk about different kinds of work that often go unnoticed and look for conditions that would facilitate them and also point to some systemic barriers that tend to make them more difficult.

To start with, the TRANFORM activities are mostly organised in what is referred to as *pilots* or around the aim to *showcase* a particular methodology or approach. However, there are significant differences in what it means to do a pilot regarding purposes and objectives, and how different partners are being involved. Also, the different forms of piloting are responding to different challenges, opportunities, and risks. The basic idea though, is usually the same: Designing an activity and *piloting* it. Piloting here implies tinkering and experimenting, but also fine-tuning a certain approach or methodology. In addition (or alternatively), a pilot can also be something – and this is more outcome-focused – that can be used as piece of evidence or showcase in a process of proofing the *added value* of RRI.

She's very sceptical about these types of processes being useful to all type of research innovation. She has always been very sceptical about this so at certain point I had managed to convince her to, I told her let's, just give us the opportunity and so let's do one small experiment and see and then it's up to you to, to judge. I mean I'm not of course trying to preach anything here. So, she said 'no let's do it the other way around. You show me the added value and if I believe in it, if I get convinced by it then I can open doors for you.'

(Int_09)

As we already indicated throughout the previous section, the different clusters conduct their RRI activities in the different regions, but they are also doing something else. As one of the cluster members told us in an informal conversation when discussing their work: "yes, but there is more than that."

Maintaining and expanding relations and mandates for action

Aside from working towards achieving the more formal objectives set out in the project's work package descriptions, the different activities in the clusters also fulfil the more informal objective of nurturing pre-existing relationships. This is the case for the work in Lombardy as well as for the work in Catalonia.

The activities of participatory agenda-setting in the Lombardy cluster are based on a broader idea about legitimate purposes and rationales of RRI work. First and foremost, this means working with decision-makers. This is expressed in the idea that for working towards more responsible R&I ecosystems you need to "work with people really involved and key in governing innovation which means not only of course policymakers but the people in charge of decision-making in the governance and in the government." (In_05)

Consequently, we see a strong focus on governance partners in the work of the Lombardy cluster and in the way our Lombardian discuss RRI in Lombardy. It is described as one mode of innovation governance in which policy- and decision-makers are the main collaborators. This focus also informs what is considered to be *good*, namely the opposite of *fake* engagement. Fake engagement in this dichotomy is a designation for engagement initiatives where there are no consequences or which are not tied in any way to ongoing policy work.

This distinction clearly resonates with work pointing to the risks of *regulatory capture* and *window dressing*. It also points to the core ambition of the work in the Lombardy cluster, which is to make sure that the input of the citizens actually matters in some way in policy- or decision-making processes. The question then becomes: How can the Lombardy cluster make sure that input from citizens is used for more than mere window dressing? Other clusters also stress the importance of being *real* about engagement. Colleagues from Catalonia refer to their citizen science project as real engagement. The idea to have *real* activities is closely linked to a model of citizen science as co-creation and a form of participatory governance.

This is where maintenance work enters the picture. Maintenance here means nurturing and maintaining pre-existing relations with administrative partners. Maintenance work is a central element of all the cluster-activities. In Lombardy, it takes the shape of collaboratively developing the engagement process and methods and it appears in activities to build capacity and develop skills within the Region. This kind of work can also be seen in efforts of making RRI principles more visible on a national level. As we have laid out in Chapter 4, our colleagues in Lombardy often talk about a *preparatory stage* as distinct from the actual engagement activities (i.e., the participatory agenda setting and the citizens' jury). This stage, in the account of our colleagues, had the purpose to make sure "that the public engagement activities were actually actionable from the Region. So, we had a lot of mutual learning meetings with [NN]" (Int_06).

The declared aim of this stage then was to increase the chance of the activities to become consequential in the policy-realm, that is to become *actionable*. To make sure that this goal is achieved, the cluster members organised several meetings that were used to define the scope as well as the purpose of these activities. In addition, these meetings were also used to build capacity, awareness, and mutual understanding.

These meetings were on the one hand geared towards improving the quality of the engagement activities. However, they were also about nurturing the relationship with partners within the Lombardy Region and making them *RRI champions* within the regional administration. This is framed as *believing* in what they are doing and trying to achieve:

> Yeah. And she's really supporting the citizen engagement activities here (laughter). It's like yeah. She really believes in this in these activities.
>
> (Int_06)

This points to a form of maintenance work that slowly expands an existing network by convincing and enrolling more and more actors. The partners within the Lombardy Region therefore act as amplifiers of the message sent via the *showcasing* activities. Without these actors within the regional administration – and thus work done before (and after) the actual TRANSFORM project – the cluster activities would have looked rather different. It is even in the realm of possibility that they would maybe not have been possible to conduct at all.

There is no question that these relationships and the practice of maintaining them are essential for the long-term success of these activities. What is often casually referred to as "preparatory" stage thus turns out to be way more than that. What our colleagues describe is careful maintenance work that builds the preconditions for the activities of the cluster. Once such relationships are established and stabilised to some degree, it becomes easier for the partners within the regional administration to work towards transformations in their own organisations. Members of the Lombardy cluster talk about these activities as a form of *dissemination* of a certain way of thinking and working together.

> So, for us at the end it's important to disseminate beyond the borders of Transform. But the real dissemination activities are within the Region and within the other departments in the three regions involved.
>
> (Int_05)

Our colleagues, however, point out that there is a certain fragility in working this way. The reliance on individual actors within the administrative ecosystem bears the risk that, in case these actors are either moved or themselves decide to move, you need to start over again. Therefore, the relationships with the Region need constant care and maintenance:

> And that's why as I said I think it's very important that we have these out-reach communication activities within the Region, within the other civil servants, with the other civil servants. We also need to plan the public, the public event to share the Transform results and also this will be very important.
>
> (Int_06)

Thus, in parallel with the work on the project activities – that is the core and most visible of the work done in TRANSFORM – the cluster members are constantly reflecting on how to best expand the network. What we see is thus a constant process of translation, also in the more classical sense of enrolment and enactment developed by Latour and Callon (Callon 1986; Latour, 1987).

> (…) it's doing things together and talking together and trying to understand each other, the Bassetti foundation with its own being a third party and at the same time being part of the cultural environment, shaping the public administrative discourse (laughter) I would say okay. And now they/and now Enza and the other, the other people there perceive their job okay and effectiveness of the job. And that the same/so being a third party and at the same time working together okay and trying to make something together.
>
> (Int_07)

The overarching objective of these activities, then, is to try and influence the "cultural environment" and also attempt to "shape the administrative dis-course." This clearly resonates with the core objective of RRI, which is to have a transformative effect on cultures of innovation governance (European Commission 2015).

The situation in the Lombardy cluster is comparable to how our colleagues from Catalonia talk about their activities. In this cluster too the activities are grounded in a strong relationship between the research partners and the regional administration. One could argue that the Generalitat of Catalonia plays an even more active role in the pilot projects in this cluster. This becomes visible in the way the so-called think tank is used. In general, the rationale for establishing think tanks in the TRANSFORM project was to engage regional stakeholders and involve them in the various cluster activities. In practice they played different roles in the clusters, and the members of the clusters were also involved to varying degrees in the project work. Our colleagues from Catalonia talk about their think tank as one of the central elements in developing the pilot projects. Interestingly, the Catalan think tank consisted of a comparatively large number of members in the beginning. The think tank was used to present and discuss citizen science as an approach for working or even governing with a group of potentially relevant stakeholders. Once the pilot projects crystallised and became more concrete, the size of the think tank decreased as the chosen projects were not relevant for all the stakeholders who initially joined the think tank. One of the main objectives beyond defining pilot projects that meet the project's assessment criteria, the objective was a cultural shift, "changing mindsets" (Int_03) as they call it.

What we think is important to stress here is that – akin to the work in Lombardy – this kind of involvement and the creation of responsiveness is conditional on work that reaches beyond the project lifetime of projects like TRANSFORM.

> And in fact, the idea at the beginning was to have much less people in the think tank. I think in the proposal you only need to have like ten people involved but because in this case, [NN] is the right key player, she was involving a lot of people. And then that's why we had so many people at the first session especially of the think tank and then the pilots were so successful. And also in Mollet and this area, she was already working there with the SeeRRI project. So yeah. They know her. Yeah. So I think in this sense, [NN] has been key, as key player in the Generalitat to involve all these people but no. It was not foreseen, no.
>
> (Int_01)

In quotes like the one above, when members of the clusters talk about the work with the think tank, there are a few recurring themes that are noteworthy. Firstly, this points again to the limitations of R&I governance through project funding and, more broadly, to the limits of what can be done in *projectified* work that has a clear start and endpoint and usually a life expectancy of three to five years. This is not new and has been noted by others before (Torka 2006).

What often gets side-lined and overlooked, though, is the constant maintenance work that creates and sustains these environments in which such success in terms of impact is possible despite the limitations of projectified modes of working towards transformation. One of the reasons why the think tanks are considered successful by the members of the Catalan cluster is the fact that there was already a group of actors – allies if you will – that were willing to join this pilot. It was this experience which led to a certain view and use of the think tank:

> I told them that if we want to have impact we need that think tank. That was, it has been a like a process. For the think tank we selected stake-holders that were already somehow engaged in the work I was doing and that could have some relation to citizen science and we open it a little bit more also.
>
> (Int_03)

The think tank in this account is used to select pilot activities. In addition, it is also a means to create and stabilise relationships with actors in the Catalan R&I ecosystem. It is conditional on previously established links and is designed to make the best use of those. In this sense, the think tank was a way to develop pilot projects and also to identify partners with an actual interest in collaborating. One administrative project partner put it like this: "once you are connected you don't need a think tank" (Int_03).

And here we see a difference to the work in the Lombardy cluster. While there the maintenance work mainly is focused on nurturing existing relations between the Bassetti Foundation and the Lombardy Region, the activities in the Catalan cluster were also geared towards bringing additional actors into the fold. The reason why the administrative partners need a project like TRANSFORM for this is – as one actor from the administration tells us – that to be able to do things differently, there always needs to be a mandate:

> So for them it's an opportunity also to start talking to each other inside the, the municipality because, yeah and to change a little bit the waste, the people that works in waste management they are somehow a little bit, well. They forgot that they are talking about people, you understand. It's like what can we do for people to convince people to do something because it's not so easy like putting containers and that so.
>
> (Int_02)

A project like TRANSFORM then is described as an *excuse* or *umbrella* to do this kind of nurturing work within the administration:

> Transform is the excuse to do that. It's an umbrella, no because this allows me to enter in many different/I have no competencies on all that, on working with universities, on working with municipalities but if I have a project I can go everywhere.
>
> (Int_03)

This form of doing something *different* has impacts beyond the regional administration and also touches on translations of citizen science among the partners in the pilot activities. One example for this are narrow understandings of citizen science as education or awareness-raising. What we see in the pilot activities of the Catalan cluster is a multi-faceted translation of RRI as citizen science:

> The doctors talking about awareness which is OK and yes I agree with them but the project is more than that. Otherwise, you just can hire a company, a media company and they will do a big campaign with a similar money amount on the problem itself, the health problem that people is not aware of. But the idea is also that without even them not be fully conscious that they're starting to navigate through this much more complex ecosystem, no.
>
> (Int_04)

Our project colleagues are very aware of such limited framings of citizen science. The crucial point here is, that they are ok with that. The idea of awareness has its place and importance in these activities, as long as there is *more than that*, as one colleague was eager to stress. This something more is precisely the kind of maintenance work that aims at creating and nurturing new links between different academic, governance, and (civil-)society actors as well as within the different institutional entities and in doing so subtly re-configuring the Catalan R&I ecosystem. In quotes like the one above actors – through their participation in TRANSFORM – start *navigating* through their ecosystem in novel ways by making new connections.

Piloting, maintenance, and power imbalances – the issue with timing

Differences in the institutional-organisational set-up within the various TRANSFORM clusters also shape how the clusters are able to make use of *windows of opportunity*.

> And when [NN] when he started being involved in the project the, the regional plan was already, well it was just published in October but you can imagine a document like that has to go through the government. So in June, May-June it was already closed basically. So there is absolutely no/I mean we can broadly say that the things we are working on are within S3 priorities. So circular economy. The topic we are touching upon is still relevant in the new regional plan. So hopefully it will inform, so what we are doing somehow will show them ways of implement S3's actions when linked with circular economy and not only hopefully in a more participatory way. But it's a bit of a stretch. I mean it's all I can say basically. It's

because it's on the topic that is rele/it's one of the S3 topics but apart from that it's well, it, it used to be because circular economy was in the previous plan. Now the, the topics are much more broader so. But there isn't basically. There is no clear connection like there is in Lombardy or like the there is more or less in Catalonia so.

(Int_09)

As we described in Chapter 6, the BCR cluster needed to develop a relationship between the different cluster members from scratch throughout the project. Consequently, it proved to be more difficult to identify so-called "windows of opportunity," moments in policy- and decision-making processes where it would have been possible – and useful – to introduce RRI principles. In the conversations we had there was some talk of opportunities that might have been and that could be developed.

However, not only did the relations between the different partners need to be built but INNOVIRIS itself also appeared to be a fragmented entity, presenting different actors and views on the collaboration throughout the duration of the project. This becomes visible in the role that a colleague from the administrative partners in the BCR cluster partly takes on when he acts as a spokesperson of TRANSFORM and RRI within INNOVIRIS. Only towards the end of the project, such a window opened up through the NoJavel! Initiative. This is because this pilot was able to link ideas about "meaningful" engagement with discussions about ex-ante project evaluation within INNOVIRIS.

In contrast, in the Lombardy cluster the collaboration between FGB and Lombardy Region is already established. This makes it easier to identify fitting windows of opportunity, but the downside of this is the risk of regulatory capture. The constellation in Lombardy makes it very easy for the Region to tell FGB where RRI fits in and where it does not. Our Lombardian colleagues talk about this as being content with small successes and being realistic (see Chapter 4).

It is thus important to note that this kind of maintenance work of identifying and making use of such windows of opportunity is premised on an already pre-existing relationship between the different partners. There needs to be some kind of trust in both the partners and the approaches they represent.

We had online meetings and we also met one time here at the foundation premises but after the, the engagement process and did the, so to build the next stage. So, the first meeting in person was one month ago I would say but we knew them already and I think as we said in some other meetings it was, I think it was very important because the connection between the Foundation and the Region was already very strong. So, there was already a lot of trust. They, they know how we work and we know how they work so.

(Int_06)

FGB, for instance, is building on decades of work on responsible modes of innovation governance, an already existing network within the regional administration and even a legal anchoring of RRI – a network that at the start of TRANSFORM was already convinced that RRI is important.

Building a network in a fragmented R&I ecosystem in the Brussels-Capital Region

The cases we discussed so far – Lombardy and Catalonia – were lucky enough to be able to build their projects on strong foundations of pre-existing working relationships and relationships of trust between the different partners. As we showed, the condition for having such relationships is continuous maintenance work, some of it behind the scenes and beyond the confines of single projects. What happens if such a basis is missing? To a certain extent, the activities and struggles we witnessed in the BCR cluster can be seen as a case in which consortium members – both on the research as well as on the administrative side – had to build such a relationship while doing meaningful project work. They had to build the ship while on sea to use a nautical metaphor.

Members from the Brussels cluster describe themselves as "totally newcomers in the scene" (Int_09). Additionally, as we already described in Chapter 6, the R&I ecosystem – in particular the governance system – is perceived as fragmented and "complex" (Int_11). Interestingly, both the research and the administrative partners see the field in this way. It can be argued, therefore, that this view is not a function of being a newcomer in the ecosystem. Furthermore, we were told that the system is very dynamic and hard to decipher. This of course makes it difficult to understand who the relevant actors are. The following episode is a good example of the difficulties the members of the BCR cluster were confronted with:

> We found ourselves in a big room with like ten people around the table (laughter). No one we knew. There was these two people from the cabinet. The person who was the director of Innoviris at the time. A guy,[NN] who used to work at Innoviris and had been moved to the government in the meantime so in that cabinet. [NN] was there. (…) And two people from the cabinet of environment. And that's when it became kind of really weird, because when I went to the meeting I thought that the discussion would be who handles the project. Because I knew the cabinet would not be able to directly work on it. But I was prepared to have other agencies involved so I took the discussion we had had in the past was that it would be either Brussels Environment which is the agency for handling all the environmental things, Hub Brussels, which is the agency focused on innovation so it's really supporting enterprises, companies and less research, more innovation oriented. (…) There is another agency that is for employment.
> (Int_09)

This little episode nicely shows how our colleague thinks of her situation within the BCR R&I ecosystem. A newcomer in a highly complex situation with numerous actors whose interests and agendas are not entirely clear. In such a context, the challenge becomes to decipher the situation and identify potential collaborators – allies if you will – and spaces for RRI-inspired work. And indeed, while setting up the process, different administrative actors were considered as potential partners for the project. Only towards the end of the process did it become clear that this partner would be INNOVIRIS. This process, however, apparently was also rather obscure or non-transparent for IN-NOVIRIS since they describe their participation in the project as "accidental" (Int_11).

In such a process, both prospective partnerships and issues have to be built at the same time. If one shifts, so does the other (and the translation of RRI together with it). Unsurprisingly, in a room like one of our colleagues described so vibrantly above, it is not at all clear what the core topics and priorities for a project like TRANSFORM will be in the end.

The most fitting adjective the BCR cluster members used to describe their situation was "disconnected" (Int_09). It should be clear by now that the situation in the BCR therefore is in some ways a contrast to how the project activities in Lombardy and Catalonia unfolded. While the latter started and then further developed from already stable relationships, our colleagues in Brussels more or less started from zero. This does not mean, however, that maintenance and care mattered less here.

What we see in this case, then, are various translations of RRI that co-emerge with the particularities of the R&I ecosystems and the place of the pilot activities within them. This also corresponds to a different form of maintenance work geared towards building trust and finding niches.

The project activities of the BCR cluster therefore needed to be more adaptive and experimenting compared to the other two clusters. The understanding that our colleagues from Lombardy talked about needed to be developed during the pilot work. In the BCR cluster, there was no person *inside* the administration who already believed in the added value of RRI principles and approaches and was willing to convince others on the inside. This trust and belief had to be built.

This was done in two ways: Firstly, by finding a pilot activity that was interesting as an RRI pilot and that was also relevant for INNOVIRIS. This was the Unsold Food case, which was relevant in terms of community development, with regard to circular economy and as a showcase for what RRI-inspired co-creation approaches could deliver for an agency like INNOVIRIS. The second approach, pushed mainly by our colleagues from UC Louvain, was to engage their university management and try to transform innovation cultures through a change in the education of future scientists and engineers. The work behind the scenes involved numerous discussions, working documents, and also workshops to find ways of productively working together by identifying problems and challenges that were relevant for all actors involved.

The overarching idea guiding both of these ways of working with the regional R&I ecosystem was to provide what is called a "protocol" that enable actors from the R&I ecosystem to assess projects and give advice on how to make them more resonant with RRI principles: The so-called "Spheres protocol."

Overall, this turned out to be a rather slow process that took several attempts at different approaches and activities and also led to some dead ends.

Summary – carriers and mediators in territorial translations of responsibility

When we look at the project activities in the different regions in comparison, we clearly see the diversity in their approaches and how they are shaped by the institutional-organisational and political contexts in which they are situation or – to phrase it in STS terminology – with which they co-emerge. It is therefore not surprising that we find diversity; however, we also want to carve out some of the things that the different clusters have in common.

Summing up this chapter we now want to point to some of the elements and mechanisms through which certain translations of RRI become stable and sustainable in the different regions: The carriers and mediators.

In their work on the translation of "technologies of participation," Linda Soneryd and Nina Amelung (2016) talk about "organizational carriers" and different "normative and symbolic systems to capture how some of these technologies become successful in getting translated into different settings. Such carriers can contribute to shaping practices and provide repertoires for interaction (Turnhout, Tuinstra, and Halffman 2019). In a similar way, Chilvers, Pallett, and Hargreaves (2018) talk about "wider spaces of participation." In describing such spaces, they provide a more nuanced picture of what different elements of organisational carriers might be. These include particular institutional-organisational settings with their ideas of who can legitimately participate in certain activities and to what extent, zones of standardisation that shape models of participation enacted through certain methods, and, finally, controversy or issue spaces, that is, the topics that can be discussed. Beyond such spaces of participation, they point to certain constitutional stabilities that shape R&I ecosystems and policies such as "laws, regulations, infrastructures, established social practices, socio-technical imaginaries, and collective forms of public reason that have become established within situated (national) political cultures over historical time." (Chilvers, Pallett, and Hargreaves 2018, 202).

The work on the Lombardy cluster can build on constitutional stabilisation as it is legally grounded in the 2016 Law "Lombardy is Research and Innovation" (Legge 29/2016 "Lombardia è Ricerca e Innovazione"), which positions RRI within the regional R&I system. This law also creates a mandate for the regional administration to build RRI capacity and to experiment with novel approaches. Still, the aim is to optimise the production capacity of regional companies.[8]

This legal framework for the work in the Lombardy cluster is the outcome of a longstanding relation between the regional government in Lombardy and FGB, as we explained in previous chapters. From its outset, RRI and the quest for *responsibility* were woven into the institutional structure of innovation governance in Lombardy on a high level. RRI thus gets translated as a variant of democratic deliberation *from above*. RRI as a participatory agenda setting is intended to help the Lombardy Region learn about the needs of the citizens so *they* can develop better policies. A citizens' jury is conducted to provide recommendations about potential social implications with regard to privacy, inclusivity, and accessibility. In all of this, the timing as well as the issues are defined in a top-down manner. This has the advantage of making sure that the RRI activities are aligned with and become responsive to what is relevant to the regional administration. The approach therefore is geared towards mitigating the risk of having *fake* participation as one colleague from Lombardy put it. This legitimate objective then means that there is no empowerment when it comes to framing the issues or deciding how the outcomes of deliberations should be used.

What is reality in Lombardy appears as an inspiration in other clusters. One of the plans of how to create a legacy of the TRANSFORM project in Catalonia is to produce standards for citizen science and push for regulations that give a mandate for more participatory forms of innovation governance. There is thus a clear awareness within the consortium that some form of institutional and even constitutional anchoring is needed to sustain the activities that were piloted in the TRANSFORM project.

It is also worth mentioning that the publication of several OECD reports on deliberative democracy (OECD 2020, 2021) plays an important role in the translation of RRI. Our colleagues talk about the effect that the publication of such reports can have in the regions:

> And of course the OECD is an intergovernmental organisation with a wide influence on policies and of course can have a role in persuading regional authorities or public authorities on investing on responsible innovation activities, because perhaps at some point in the future they will be obliged to comply with these recommendations.
>
> (Int_05)

Intergovernmental institutions like OECD, however, have an influence that goes beyond assisting regional actors in "persuading" local authorities to embrace certain modes of governance. They can be a core element in a particular network and thus shape certain translations of RRI. The push of deliberative democracy by the OECD – which also was embraced by the European Commission and DG Joint Research Centre with its site in Lombardy – likely made it easier to convince the Lombardy Region about the added value of citizens' juries.

When talking about organisational carriers it is also important to highlight that this is often about the support of single departments within public administration institutions. In some cases, there are single persons who push for the integration of RRI principles or the experimentation with more deliberative and participatory methods. We described in this chapter what the lack of such support means for a project like TRANSFORM.

Finally, as we argued throughout the empirical descriptions of the different regional clusters as well as in these comparative chapters, a crucial element in the territorial translations of RRI are the methods through which RRI gets enacted. There is participatory agenda setting and citizens' juries under the broader umbrella of deliberative democracy in Lombardy, citizen science and shared agendas in Catalonia, and co-creation and (to a lesser extent) design thinking in the BCR.

These methods help translate RRI into something "concrete" and in doing so it helps to relate constitutional and institutional stabilities to the lived practices of doing innovation governance. As one colleague put it, this is about *providing something concrete*, about making an abstract principle tangible.

> Maybe the, so we are strong. We as Lombardy, I think we are strong enough under the legal framework, under the legal point of view to embed RRI in our legal system. Transform is now I think is providing (…) let's say a concrete/it's something concrete that shows to [NN] and the rest of the group that responsible innovation is more than a principle. And okay but everybody knows. Okay so but okay but then what is it?
>
> (Int_07)

But having these kinds of mediators also gives a particular shape to RRI and enacts a certain version of responsibility. This can be responsibility for making the inputs of citizens count, but it can also mean giving the citizens power over what they are allowed to discuss and how exactly their inputs are being used. Responsibility can also be enacted on different scales through the choice of different methods: From the patient level in one of the Catalan pilots to the community level in Brussels (and also in Catalonia) and even the national level (in the case of the translation of the OECD report).

And finally, it is important to point out how the overarching framing of these methodological mediators factors into the translation of RRI. It makes a difference if there is experimenting and tinkering with RRI principles and approaches or if these methods are applied for showcasing and proofing as is mostly the case in the TRANSFORM activities. That they are generally referred to as "pilots" by our colleagues underscores this mode of showcasing something (mostly some sort of added value for administrative actors).

We have shown in this chapter that there is a fascinating diversity when it comes to the translation of responsibility in the three TRANSFORM clusters. These translations become visible, for example, in different ideas about good engagement, the show up as distinct practices of assuring a legacy and impact of the cluster activities, which themselves co-emerge with certain models of impact pathways, and they can be observed in various forms of maintenance work that provide that grounds on which the cluster activities can thrive.

In previous chapters, we have also argued that the emergence of RRI as a policy term can be understood as a translation as the concept managed to bring together different strands of technology assessment, ELSI/ELSA research with academic disciplines like STS and ethics as well as with policymakers from different national and transnational administrative institutions.

In the next chapter, we will compare such translations of responsibility on a national and international level.

Notes

1 The challenges of such a framing are discussed among SwafS-funded RRI projects. One instance of this is the discussion paper "Impacts, Pathways and Benefits of RRI" authored by members of the Super MoRRI project consortium as a part of this project. https://super-morri.eu/findings/. Accessed March 3, 2023.
2 https://www.transform-project.eu/wp-content/uploads/2022/11/Report_TRANS-FORM_CitJury.pdf. Citizens' Jury on Responsible Smart Mobility. Report of the deliberative process conducted within the EU H2020 TRANSFORM project. Accessed December 20, 2022.
3 Ibid.
4 Ibid.
5 https://www.transform-project.eu/wp-content/uploads/2022/11/Roadmap_Lombardy Cluster.pdf. Accessed March 2, 2023.
6 https://www.transform-project.eu/catalonia/. Accessed February 15, 2023.
7 A condensed version of this section was by submitted by the authors of this book as a part of the empirical analysis in the paper "Transformative Translations? Challenges and Tensions in Territorial Innovation Governance" to the journal Novation: Critical Studies of Innovation. The manuscript was accepted for publication pending minor revisions in March 2023.
8 The original text reads: "la cultura della ricerca e dell'innovazione responsabile, anche attraverso la diffusione della conoscenza nel tessuto imprenditoriale lombardo, la sperimentazione e la divulgazione di metodi e processi innovativi finalizzati a ottimizzare la capacità produttiva delle imprese operanti in settori tradizionali" (LR 29/2016 "Lombardia è Ricerca e Innovazione").

References

Arora, Saurabh, Barbara Van Dyck, Divya Sharma, and Andy Stirling. 2020. "Control, Care, and Conviviality in the Politics of Technology for Sustainability." *Sustainability: Science, Practice, and Policy* 16 (1): 247–62. https://doi.org/10.1080/15487733.2020.1816687.

Bensaude Vincent, Bernadette. 2014. "The Politics of Buzzwords at the Interface of Technoscience, Market and Society: The Case of 'Public Engagement in Science.'" *Public Understanding of Science* 23 (3): 238–53. https://doi.org/10.1177/0963662513515371.

Callon, Michel. 1986. "Some Elements of a Sociology of Translation: Domestication of the Scallops and the Fishermen of St. Brieuc Bay." In *Power, Action and Belief. A New Sociology of Knowledge?* edited by John Law, 196–233. London/Boston/Henley: Routledge/Kegan Paul.

———, and Bruno Latour. 1981. "Unscrewing the Big Leviathan: How Actors Macro-Structure Reality and How Sociologists Help Them to Do So." In *Advances in Social Theory and Methodology: Toward an Integration of Micro- and Macro-Sociologies*, edited by Karin Knorr Cetina and Aaron V Cicourel, 277–303. Boston: Routledge and Kegan Paul. https://doi.org/10.4324/9781315763880.

Chilvers, Jason, Helen Pallett, and Tom Hargreaves. 2018. "Ecologies of Participation in Socio-Technical Change: The Case of Energy System Transitions." *Energy Research and Social Science* 42 (August): 199–210. https://doi.org/10.1016/j.erss.2018.03.020.

European Commission, Directorate-General for Research and Innovation. 2015. "Indicators for Promoting and Monitoring Responsible Research and Innovation: Report from the Expert Group on Policy Indicators for Responsible Research and Innovation." Publications Office. https://data.europa.eu/doi/10.2777/9742.

Guimarães Pereira, Ângela. 2015. *Science, Philosophy and Sustainability*. Routledge. https://doi.org/10.4324/9781315757902.

Jasanoff, Sheila. 2003. "Technologies of Humility: Citizen Participation in Governing Science." *Minerva* 41 (3): 223–44. https://doi.org/10.1023/A:1025557512320.

———. 2005. *Designs on Nature: Science and Democracy in Europe and United States*. Princeton: Princeton University Press.

———, and Sang-Hyun Kim. 2015. *Dreamscapes of Modernity: Sociotechnical Imaginaries and the Fabrication of Power*. Chicago: University of Chicago Press.

Kjølberg, Kamilla Lein, and Roger Strand. 2011. "Conversations about Responsible Nanoresearch." *NanoEthics* 5 (1): 99–113. https://doi.org/10.1007/s11569-011-0114-2.

Latour, Bruno. 1987. *Science in Action. How to Follow Scientists and Engineers Through Society*. Cambridge: Harvard University Press.

OECD. 2020. *Innovative Citizen Participation and New Democratic Institutions. Catching the Deliberative Wave*. Paris: OECD Publishing. https://doi.org/10.1787/339306da-en.

———. 2021. *Evaluation Guidelines for Representative Deliberative Processes*. Paris: OECD Publishing. https://doi.org/10.1787/10CCBFCB-EN.

Owen, Richard, Phil Macnaghten, and Jack Stilgoe. 2012. "Responsible Research and Innovation: From Science in Society to Science for Society, with Society." *Science and Public Policy* 39 (6): 751–60. https://doi.org/10.1016/j.respol.2013.05.008.

Puig de la Bellacasa, Maria. 2011. "Matters of Care in Technoscience: Assembling Neglected Things." *Social Studies of Science* 41 (1): 85–106. https://doi.org/10.1177/0306312710380301.

Schatzki, Theodore R, Karin Knorr Cetina, and Eike von Savigny. 2005. *The Practice Turn in Contemporary Theory*. London and New York: Routledge.

Shove, Elizabeth. 2010. "Beyond the ABC: Climate Change Policy and Theories of Social Change." *Environment and Planning A* 42 (6): 1273–85. https://doi.org/10.1068/a42282.

Soneryd, Linda. 2015. "Technologies of Participation and the Making of Technologized Futures." In *Remaking Participation: Science, Environment and Emergent Publics*, edited by Jason Chilvers and Matthew Kearnes, 144–61. London and New York: Routledge.

―――, and Nina Amelung. 2016. "Translating Participation: Scenario Workshops and Citizens' Juries Across Situations and Contexts." In *Knowing Governance. The Epistemic Construction of Political Order*, edited by Jan-Peter Voß and Richard Freeman, 155–74. UK: Palgrave Macmillan. https://doi.org/10.1057/9781137514509_7.

Stilgoe, Jack, Richard Owen, and Phil Macnaghten. 2013. "Developing a Framework for Responsible Innovation." *Research Policy* 42 (9): 1568–80. https://doi.org/10.1016/j.respol.2013.05.008.

Strand, Roger, and Jack Spaapen. 2021. "Locomotive Breath? Post Festum Reflections on the EC Expert Group on Policy Indicators for Responsible Research and Innovation." In: *Assessment of Responsible Innovation. Methods and Practices*, edited by Emad Yaghmaei and Ibo van de Poel, 42–59. Routledge. https://doi.org/10.4324/9780429298998.

Torka, Marc. 2006. "Die Projektförmigkeit der Forschung." *Die Hochschule: Journal für Wissenschaft und Bildung* 15 (1): 63–83. https://doi.org/10.25656/01:16427.

Turnhout, Esther, Willemijn Tuinstra, and Willem Halffman. 2019. *Environmental Expertise*. Cambridge: Cambridge University Press. https://doi.org/10.1017/9781316162514.

Vinsel, Lee, and Andrew L. Russell. 2020. *The Innovation Delusion : How Our Obsession with the New Has Disrupted the Work That Matters Most*. New York: Random House.

von Schomberg, Rene. 2012. "Prospects for Technology Assessment in a framework of responsible research and innovation." In: *Technikfolgen abschätzen lehren: Bildungspotenziale transdisziplinärer Methode*, 39–61, Wiesbaden: Springer VS.

8   A sideview to the laboratory

Introduction: Our case-control

With this book we wish to develop insights and lessons about responsible governance of technoscience mainly by pursuing a project in which the icons of technoscience – laboratories, white coats, reagent tubes, and expensive measuring devices – were almost out of view. In our own reflection process, we could not help contrasting the case(s) of TRANSFORM with the other responsible research and innovation (RRI) projects we were involved in, which at least in our Norwegian context frequently meant projects designed to "implement" RRI in academic research within emerging sciences and technologies such as bio- and nanotechnology and informatics. To develop our arguments about translations of RRI in TRANSFORM and broader, we have included this little "sideview" chapter, a chapter that compares and contrasts TRANSFORM with our experiences in an explicitly technoscientific endeavour, namely the Centre for Digital Life Norway (often abbreviated DLN) and in one of its associated projects, the Centre for Cancer Biomarkers at the University of Bergen (CCBIO).[1] In what follows, we shall first provide a short introduction to what DLN was (and still is at the time of writing, 2023) and what translations of RRI we have observed within DLN. We then move to CCBIO to get even closer to the laboratories. Finally, we discuss our TRANSFORM experience with DLN and CCBIO as contrasts.

The Centre for Digital Life Norway (DLN)

In 2014, the Research Council of Norway (RCN) published a white paper called *Digital Life – Convergence for Innovation* (RCN 2014). The paper was followed up by several calls for proposals that led to the creation of a national centre for "digital" biotechnological research as well as a portfolio of research projects across Norway that were designated member projects of the centre. Of relevance to this book, RRI was from the onset identified as a horizontal, crosscutting principle of the centre, to be implemented in all activities including the

DOI: 10.4324/9781003371229-8

individual research projects themselves. The centre was called Centre for DLN. It was operative by 2015 and is now in its second funding period, 2021–2026.

Norway is not a member state of the European Union (EU) but takes part in most EU activities and developments as part of the so-called European Economic Area. Norway is a full member of the European Research Area and participates in the EU framework programmes for research and development/ innovation. While the country in principle is autonomous and sovereign with respect to its research and innovation (R&I) policy, policymaking in practice is heavily influenced by the EU and the OECD. Notably, Norway has been following policy trends of supporting generic technologies and the concept of "challenges." While the top-level R&I policy is anchored in the various Ministries, in particular the Ministry of Education and Research,[2] and ultimately in the Norwegian parliament, the main public research funding organisation is the RCN. In terms of institutional hierarchy, the RCN is close to being a directorate under the Ministry of Education and Research. It was formed in 1993 as the result of a merger between several field-specific public bodies with a mandate to fund research. The grant handed out by the RCN is the main source of income for the Norwegian sector of research institutes outside of universities. For the university sector, RCN is the largest source of so-called "external funding," meaning research funding other than what is already included in the direct public budget of the universities, such as salaries for permanent positions. All of this is to say that RCN is an important actor in Norwegian research as a funding organisation. In addition, it is an important policy actor both in terms of its own grant policies and in the capacity of being an advisor to the ministries, in particular the Ministry of Education and Research. In its various capacities, a core ambition has been to facilitate and enact change agency for Norwegian R&I. Specifically, the RCN developed a number of policies that aimed to change the direction of Norwegian research by developing new infrastructure, new collaborations, and new foci of research. The role of RCN as a change agent has not been without controversy; over the years, several university rectors have expressed dismay at the idea that the research council knows better than universities about their needs to change.

Not the least, the RCN has had a keen interest in supporting and directing Norwegian life science into the so-called post-genomic era. In the period 2002–2011, one of its main funding programmes was called "FUGE" – FUnctional GEnomics in Norway. Through its calls, FUGE encouraged Norwegian research life science environments to collaborate more and also coordinate a division of labour with respect to new technological infrastructures. Furthermore, FUGE gradually incorporated ever stronger ethical, legal and soci(et)al aspects (ELSA) dimensions into the programme with an orientation that was not too unlike the FP7 Science-in-Society (Nydal 2006). The sequel to the FUGE programme operated from 2012 to 2021 and was called BIOTEK2021; indeed, it was not the only funding programme named "—2021," with obvious reference to Horizon 2020. In its first call, BIOTEK2021 funded a set of large research projects (by Norwegian standard, with a budget of approximately 4M€ per project); for these projects, the inclusion and integration of ELSA was a mandatory requirement. By 2014, the RCN replaced the ELSA concept with that of RRI, looking explicitly to the EU but even more to the UK, to the Engineering & Physical Sciences Research Counciland its anticipate, reflect, engage and act (AREA) framework for responsible innovation (Gulbrandsen 2022).

In this light, the content of the RCN white paper on *Digital Life* hardly contained any surprises. It combined the policy goal of supporting emerging technologies with that of pursuing societal challenges and invoked RRI, inter- and transdisciplinarity to that effect. Furthermore, *Digital Life* addressed what was perceived as a particular limitation in Norwegian life science, namely that biologists and biotechnologists did not have sufficient collaboration with the "hard" natural sciences such as mathematics, informatics, physics, and engineering. The recommendations of the white paper were adopted, and in 2015 DLN was created as a "national virtual centre" for biotech R&I. In the first place, this involved a national networking project with a scientific director and a team of managers who were responsible for the various tasks of the project – communication, RRI, digital infrastructures, innovation advice, and setting up a research school – each under the supervision of appointed academic leaders. Most of this predominantly administrative-technical support structure was located in Trondheim, at the Norwegian University of Science and Technology, but also with offices at the University of Bergen and the University of Oslo.[3]

At the same time, six large-scale (again, by Norwegian standards) research projects were funded through the first *Digital Life* call for proposals. The call text was complex and asked for a lot: Documented scientific excellence at the international level; potential for innovation and "creation of value" for Norwegian society; interdisciplinary collaboration between "wet lab" life science and "hard," mathematical or digital science; and the inclusion and integration of RRI into the project. While the national networking project had designated RRI staff, they were envisaged more of a coordinating and advisory function. By agreement between DLN and the RCN, the research projects themselves and their principal investigators were responsible for putting RRI into action.

Over the years, DLN has continued to grow. Later calls for research projects almost tripled the portfolio of DLN member projects, and the centre also opened up for other, existing research projects in Norway to become non-funded "associated projects." Such associated projects would have to show that they were "digital" biotech projects, that is, somehow combined life science with mathematical modelling, bioinformatics or other "hard" science. In theory, they were also required to include an RRI dimension; in practice, however, the bar was set low for that requirement. The research school created a set of courses and other activities for PhD candidates and early career scientists at the postdoctoral level and counts with more than 400 members as per 2023. The networking project has also continued to expand its services, including workshops, annual conferences, individual advice, innovation roadmaps, and seed funding especially directed towards innovation initiatives at low levels of technological readiness. There have also been some incentives and initiatives to create research collaborations across the projects in the centre. Their impact has not been systemically assessed. On one hand, there have been few formal collaborations at the senior, "PI" level. On the other hand, the impact of small-scale cross-project, cross-city collaborations between early career researchers may turn out to be significant in the long run.

For being a centre, then, DLN became a rather decentralised operation, more a loose but still formalised network of research projects that got the additional benefits from a well-funded research school and other services and infrastructures through the networking project. Indeed, it has been a continued challenge when presenting DLN to outsiders to explain what it is, and even why it is called a centre. To the extent there is a "DLN identity," this is perhaps more pronounced in the group of early career researchers who did their PhDs within DLN (Hesjedal 2022).

One issue within the DLN itself has been the extent to which the research that takes place within the centre, can be said to be transdisciplinary. The original white paper emphasised transdisciplinarity as the way forward but had no consistent definition or usage of the term. The same can be said for the entire discourse in and around DLN, in the centre itself and at the research council. Analyses concluded that one had been saying "trans" and essentially had been having in mind what academic literature mostly would call "interdisciplinarity" (Hesjedal and Åm 2022). Knowledge production in the research projects has to a small degree involved non-academic knowledge contributors. Interestingly, this criticism is well-known as it has been performed at several occasions within DLN, at events as well as in a designated white paper (Hesjedal and Strand 2021). Still, the term "transdisciplinary" continues to be used more or less synonymous with "cross-disciplinary" as can be verified by the DLN website.

Translations of RRI in DLN

In a speech during World War II, in 1942, President Franklin D. Roosevelt exclaimed: "Look to Norway!" There is nothing Norwegians enjoy more than being referred to as good examples. Indeed, in informal discussions around Europe, we have experienced more than once that the RCN and DLN have been invoked as RRI champions or even lighthouses. Of particular significance, Digital Life project proposals were not only required to include RRI. They were also *scored* by international peer review on the quality of their RRI plans. We may recall from Chapter 3 that Horizon 2020 never got to mainstream RRI outside of SwafS. RRI was mentioned in some calls as a relevant cross-cutting principle, but there was no explicit scoring of RRI in the evaluation procedure.

Furthermore, the networking project has throughout its existence included RRI staff equivalent to one full position and one to two part-time positions, in addition to RRI staff in the various member research projects.

As mentioned, within the RCN, RRI was seen as a continuation and a development from integrated ELSA/post-ELSA, also with a sideview to CSR, corporate social responsibility. Furthermore, among Elisabeth Gulbrandsen (2022) and her colleagues at the RCN who crafted the national RRI policy within the council, RRI was invoked from a theoretical foundation in third-generation R&I policy. RRI was seen as one of the available approaches to the perceived transformation failure of R&I systems; transdisciplinarity was considered another. Ambitions were high: By means of transdisciplinarity and RRI, biotechnology research in DLN would become transformed in a way that would "create value for Norway," where value was to be taken in a broad sense, for the economy, for society, and for sustainability and the natural environment. DLN, which arguably did not amount to more than one per cent of Norwegian biotechnology research, was in this way explicitly imagined to be a lighthouse that would shine and show the way into the (good) future. This is already an important facet of the translation of RRI, transdisciplinarity, and the transformative ambition: It should take place in a way that would be compelling to non-RRI, non-transdisciplinary, and non-transformed actors in the R&I ecosystem. Why would they otherwise consider DLN a lighthouse?

As is usual with experiments recruiting social sciences and humanities scholars, there is once again no paucity of chronicles. In the case of Norway and the introduction of RRI in the RCN, perhaps exactly because of the enormous ambition, the chronicles/research papers have the characteristic flavour of disappointment and insights into why the ambitions were not fulfilled. Åm (2019) showed how scientists "try to accommodate rather than enact ELSA and RRI." Delgado and Åm (2018) discussed some of the tensions present in RRI-like collaborations. As mentioned before, Hesjedal and Åm (2022) critically discussed the (non-)enactment of transdisciplinarity, while Solbu (2021) made similar observations around the innovation initiatives within DLN, see also Hesjedal (2022). Much of this research was part of or associated with DLN's own concomitant action research project *Res Publica*. Borch and Throne-Holst (2021) were RRI scholars who participated in one of the DLN member projects and came to a rather disappointed conclusion:

> If RRI had been applied from the very beginning of the project period, the chance of realising proof of concept within the scheduled time may decrease. The researchers' solution to this dilemma was to prioritize proof of concept and postpone RRI activities to later stages of the project. If RRI is expected to live up to its ambition of representing a new way of doing science, more effort is needed at the political level to facilitate change.
>
> (p. 1)

All of this sets the stage for what is the concern of this book, which is not the assessment or evaluation of RRI activities or initiatives as judged from the perspective of an ambition or a given definition of what RRI is and ought to be, but rather the empirical question of what form how such activities and initiatives come to take. We do not pretend to have a full overview to answer that question but we can outline main activities in sufficient detail for the contrast to be made with TRANSFORM.

Firstly, DLN RRI was institutionalised as a task in the networking project, with its own dedicated staff. In the first funding period of DLN (2015–2021), there was a designated academic leader for RRI within DLN who supervised one full-time and up to two part-time RRI coordinators. As an indication of the proximity and entanglement between action and scholarship, and also in the name of transparency, we name these leaders: The first one was Heidrun Åm who was also the PI of the Res Publica project, and her successor was our co-author Roger Strand. In the second period (2021–2026) the structure was adjusted, with one to two hired RRI coordinators supplemented by a part-time adjunct professor (still the same Strand).

What kind of work was this staff involved in, and what translations of RRI did that work amount to? We can name a few:

- **Teaching.** RRI was something that was taught to the PhD students of DLN in a specific "RRI course" (Hesjedal et al. 2020) as well as in minor workshops both for junior and senior researchers within DLN. At the same time, there was a common understanding among the teachers that they wanted to avoid prescriptive "preaching" of RRI principles and rather tried to inform and engage the predominantly science, technology, engineering and mathematics-(STEM)-educated PhD students in the basics of science and technology studies. In the minor workshops, however, RRI principles had to be presented because researchers knew that RRI was a mandatory requirement and hence they wanted some information about it.
- **Grant proposal writing.** While the RRI networking project staff was not directly involved in grant proposal writing (to our knowledge), they offered mini-workshops also to applicants to the Digital Life calls. The noteworthy observation in this respect is the need for these workshops: The research proposals were expected to include RRI, but few of the principal investigators were familiar with the concept. Furthermore, other than referring to quite short policy documents that focused on the justification for RRI and in particular the AREA framework, the research calls did not provide details on how RRI should be included and integrated into the projects. Part of the work of the DLN RRI staff was to try to give advice on this question, which was no easy task because the staff itself was uncertain about how reviewers were going to score the RRI plans.
- **Advice and coordination of project RRI efforts.** RRI staff in the networking project liaised with RRI staff in research projects and gave occasional

advice. In the initial years, the staff also tried to map the state of play of RRI in the individual projects; however, this was met with resistance from some biotech researchers as it was seen as unwelcome and undue policing.

- **Taking part in the governance of DLN.** A large, if not the largest, part of the workdays for RRI staff, however, was to take part in the general activities of the DLN networking project. This included smaller and larger events such as conferences and meetings, but also the day-to-day management decisions together with the director and other coordinators. RRI work was also simply to participate in the general governance of the centre. Indeed, RRI staff was to a considerable extent involved in liaising with the research council, revising the strategy for the centre, designing its midterm evaluation procedure, etc. A possible explanation for this was their experience not only with RRI but the field of R&I policy in general. RRI staff knew what policy work was and knew how to speak and write at a strategic level. Also, they could (at least try to) act as mediators between the third-generation R&I policy discourse within the RCN and DLN staff who were less familiar with policy concepts.

- **Having tensions within DLN.** Perhaps often thought of as frustration, the experience and enactment of tensions and disagreements can also be seen as RRI work (Strand 2019). One of us, having worked in the DLN networking project, can confirm that such work also took place there. Borch and Throne-Holst (2021) formulated one of these tensions, namely between expediency and RRI. It is demanding to make innovation happen in the first place, and RRI could be experienced as an additional demand. Moreover, science communication services often work within a logic by which the bright sides of what one has to offer is what one wants to display. The AREA framework, on the other hand, asks even more what may be go wrong and what possible adverse effects there may be. Such tensions were enacted, negotiated, and to some extent resolved.

We do not know too much about what happened in the research projects that were required to include and integrate RRI. Some of them employed designated RRI staff, for example, a PhD student or (part-time) postdoctoral researcher from the social sciences or humanities, supposedly to "deal with" RRI. Some of the cited work – by Hesjedal, Borch and Throne-Holst – are examples of research outputs from "RRI persons" within the DLN biotechnology projects. It would be quite implausible, however, to see these research papers as contributions to the biotechnology projects themselves and as part of an integrated, transdisciplinary effort to "create value." Rather, they are SSH contributions to RRI scholarship. This point deserves a moment of reflection in order not to be missed: In STEM projects that hire SSH researchers to "do RRI," one significant translation of RRI is *scholarly texts on RRI*, mainly written for other SSH researchers who read and write scholarly texts on RRI. One might expect, however, that both the process of such RRI research – including fieldwork and research interviews – might be conducive of anticipation and reflection processes within the STEM environments and their research projects, along the lines of the idea of RRI as "midstream modulation" (Fisher, Mahajan, and Mitcham, 2006).

To what extent did midstream modulation take place? Did the presence of "RRI persons" lead to changed practices in the biotech laboratories or in the bio-informatics and mathematics departments? And did researchers within the DLN biotechnology projects come to experience that RRI was integrated, not as an external add-on, but as an aspect of what then would become responsible digital life research? We do not know. What we can say is that we have seen no indications thereof in STEM research outputs from these projects. For one, these outputs do not mention RRI. Moreover, we do not know of RRI scholarships in and around DLN that make claims to this effect. Obviously, the absence of evidence should not be equated with evidence of absence, and it may also be too early to know. What is known, is that Hesjedal performed a set of interviews with a sample of DLN researchers and could not see indications of RRI integration:

> Though often described as rewarding, having 'RRI people' in the project was experienced as time-consuming, taking time from the 'real science'. This was a recurring point, though problematized in one of the discussions where two senior researchers discussed the challenge of allocating time for RRI:
>
> RESEARCHER 1: *But the question is: Is the solution, then, to set aside time, we always talk about time here, right, as if this was something that we lose when we do something, but, but if it's part of the culture, then it doesn't, then it immediately isn't a discussion of how many hours do use for it, but it's just part of the everyday thinking of how you do things, right?*
> RESEARCHER 2: *Yeah.*
> RESEARCHER 1: *So ...*
> RESEARCHER 2: *That's the culture ...*
> RESEARCHER 1: *That's the culture, not the time, right?*
> RESEARCHER 2: *But we establish that culture, perhaps, you said that there are no incentives for doing that. I would rather, then, have my post-docs or PhD students just do those extra experiments rather than talk to these social scientist guys.*
>
> The authors of this report include this excerpt because it is quite representative of our experience of discussing such topics within (and outside of) DLN.
>
> (Hesjedal and Strand 2021, 14–15)

Along similar lines, Aasheim (manuscript) observed in her fieldwork from a DLN project how thematisation of RRI at a project meeting could lead to a seemingly confused discussion about the nature and purpose of RRI, which at the same time developed into a conversation that *was* RRI in the sense that it had strong anticipatory and reflexive dimensions. Having "RRI people" and having such conversations could be appreciated as rewarding but perhaps most of the time something different and external, except in rare occasions where some social-epistemological insight emerged in a way that was relevant to the research itself.

The Centre for Cancer Biomarkers (CCBIO)

The evidence presented for translations of RRI in DLN still remains at some distance from the white coats and the PCR machines. Did RRI govern technoscience, or did it merely govern the governance of technoscience, if at all? We could not give a definitive answer because we did not observe or interact with the DLN member projects and it is still too early to review the research of those who did.

The home affiliation of the authors of this book – the Centre for the Study of the Sciences and the Humanities at the University of Bergen[4] – has, however, enjoyed a close relationship with one of the associated DLN projects from its very beginning, namely the Centre for Cancer Biomarkers, CCBIO.[5] To continue with our case control, we shall accordingly zoom into that project and outline the translations of RRI that we have experienced there as ELSA/RRI action researchers. Again, the experiences belong mainly to Strand, and the narrative presented builds strongly on their previous work (Blanchard 2016; Blanchard and Strand 2018; Bremer and Strand 2022).

CCBIO came into existence in 2013 when it received an RCN grant to become a Centre of Excellence. These grants are prestigious and (again, by Norwegian standards) huge. CCBIO received a total funding of 170 MNOK (approximately 17 million euros) over a total of 10 years, in addition to substantial in-kind contributions from the host university. The funding programme was located within the Science Division of the RCN and as such its policy justification narrative was more one of "scientific excellence" than of challenges or transformation. Still, as pointed out by Hellström (2018), a common feature of centre of excellence funding is that documented excellence is an *ex ante* condition and accordingly can hardly be the single goal of the enterprise. Rather than simply asking (and paying) for "more excellence," such funding programmes tend to have a more or less explicit subtext of promoting some kind of transformation of the national research capacity in the field, for instance by creating critical mass for new collaborations or medium-term opportunities to pursue research avenues that otherwise would have been difficult to pursue. CCBIO was above all a case of critical mass for new collaborations, bringing together strong but rather small groups of cancer researchers in Bergen that previously had not interacted very much.

A returning issue around RRI and ELSA as soft governance measures in STEM research is the question of whether such measures should be voluntary or mandatory. Delgado (2013) presented strong arguments against mandatory requirements in a policy brief to the RCN itself. As we have seen above, however, DLN and indeed BIOTEK2021 opted for mandatory RRI. CCBIO was different in this sense because it was as a Centre of Excellence, with no other ties to it than its proposal to perform its own, researcher-initiated research. For CCBIO, ELSA and later RRI was entirely voluntary. The initiator and later Scientific Director of the centre, Lars A. Akslen, contacted social scientists and humanities scholars at the University of Bergen at the proposal stage and invited them to include what was originally conceived as ethics and economics research as part of the centre of excellence. His reason was that new cancer biomarkers, in addition to health benefits, potentially could give rise to new ethical dilemmas as well as challenges related to cost-effectiveness. For example, biomarkers might end up becoming grounds for denying a specific treatment to certain patient groups that otherwise would have received it. This could happen if an expensive treatment was shown to be less effective in a patient group to the extent that the (economic) cost to (health) benefit ratio exceeded regulatory thresholds in healthcare systems.

Research on such dilemmas did take place within CCBIO, mainly by the recruited medical ethicists and health economists. The STS researchers within ethics line of research, however, took a broader ELSA approach (Blanchard 2016), and during the first years of the centre, the ELSA research was reconceived as RRI. Indeed, the Director himself co-authored an opinion piece that subscribed to the AREA conception of RRI as a guiding framework for cancer research (Strand and Akslen 2017).

Translations of RRI in CCBIO

We shall proceed to describe how was RRI enacted and translated in CCBIO. Both CCBIO and DLN were (and are) *centres*, though quite different ones. DLN is a highly virtual centre in the sense that is a loose network of research projects together with a networking project that does not perform research itself but offers various services as well as engages in interactions with the projects, the RCN, and various other stakeholders. CCBIO is also a virtual centre in the sense that its different research groups are not located in one big laboratory. Most of the groups are physically quite close though, in the same city and at the same campus, and most of them interact with several other groups on a regular basis. In some ways, including in the formal, administrative sense, CCBIO is best seen as one huge research project. Despite these differences, we shall see that there are many similarities between the translations of RRI in the two centres.

First of all, although RRI was embraced in the opinion piece by the CCBIO director, RRI by and large remained something that designated "RRI persons" did, above all the mentioned Blanchard and Strand and some more students and colleagues. There are exceptions and we shall return to them. However, the big picture is that RRI in CCBIO amounted to what these RRI persons did and the impacts of what they did. This includes:

- **Teaching.** A designated two-week PhD course with the title "Cancer Research: Ethical, Economic and Social Aspects" was developed and taught on a regular basis. It included introductions to STS, ethics, philosophy of medicine and health economics, and it engaged the students in reflexive, critical discussions of their own research projects during the classes and in the essays that they were required to write after the course. In otherwise rather hierarchical research environments, the course was designed to be a "safe space" where young researchers could express and discuss doubts and concerns about the projects they were involved in, very much along the Anticipate and Reflect axes of the AREA framework. A structural challenge, however, was that without the senior researchers involved, there was often little that could be done in terms of responding to the doubts and concerns raised. However, the course also served as RRI capacity building in the environment as such, in ways very similar to the equivalent course in DLN. In some cases, participation in the course developed into larger reflexive undertakings that had impact on PhD candidates' further research trajectory. Notably, seven of the co-authors in the CCBIO anthologies within the field (Blanchard and Strand 2018; Bremer and Strand 2022) had been students in this course.
- **RRI and ELSA research.** The RRI persons performed and were expected to perform research, which fed into ELSA and RRI scholarship, such as the present book. Indeed, for a centre with excellent funding, publishing was important. However, most of the publications among the SSH scholars in CCBIO were directed to at least a hybrid audience and written in formats that in principle were accessible to (medical) cancer researchers. The extent to which these outputs were read by the other CCBIO researchers – in spite of wide distribution – is at best uncertain. Our experience was that our publications were not much discussed and that the uptake was low, along the lines of Hesjedal's observation: They had little time for such things.
- **Slow emergence of interdisciplinary research.** "RRI persons" formed collaborations with medical researchers that little by little grew into interdisciplinary research crossing the "two cultures" between STEM and SSH. Examples include conceptual research on biomarker quality (Blanchard and Wik 2018; Bremer, Wik, and Akslen 2022) and phenomenological research integrated into a clinical trial (Gissum et al. 2022).
- **Taking part in the daily life of the centre and its governance.** Similar to the RRI staff in the DLN networking project, the "RRI persons" in CCBIO

spent considerable time attending seminars, workshops for early career researchers, meetings for principal investigators, etc. This was not thought of as fieldwork but rather as part of the work itself. Occasionally they said something that perhaps was an "RRI point" and sometimes it could be met with the tensions described in the DLN case. More than often, however, there was no very sophisticated RRI point to be made and they participated perhaps more as a sort of academic laypersons.

Just as with DLN, "RRI persons" in CCBIO – in particular Strand – got increasingly involved in the governance of the centre, for example, by contributing to midterm reports and final reporting. Perhaps this was a matter of having knowledge about the world of R&I policy and governance. Another possible explanation is that in his capacity as an "RRI person" his stakes in the majority of decisions were smaller than those of the medical principal investigator. Both in DLN and CCBIO one can see the contours of an RRI persona that is a "critical friend" who contributes with administrative and strategic support. Responsibility is enacted, but it is not so specifically "RRI" in the von Schomberg sense of being a fourth hurdle and a remedy to the Collingridge dilemma. It is more the everyday responsibility of accountability and care.

Even if RRI got closer to the laboratory benchtop in CCBIO than what we have evidence of in DLN, the overall picture is one of few signs of real integration of RRI into the STEM research, of creating Responsible Cancer Research in the sense of the AREA framework. Some exceptional developments took place, though. What we called slow emergence of interdisciplinary research belongs to that category. The most exceptional case, however, was the change in the career trajectory of one early career scientist who began with taking the RRI PhD course. Her own narrative is included in the textbox below. To summarise, she combined her biomedical cancer research with reflexive critical work on this line of biomedical research by means of STS and philosophy of medicine. This led to publications and conference papers on ELSA and RRI issues and even the integration of STS concepts such as sociotechnical imaginaries in the PhD dissertation itself, which was a fairly technical piece on genes involved in leukaemia (Engen 2020). Following the PhD, she became an "RRI person" herself in another large-scale biomedical research project as post-doctoral research fellow.

I have always been interested in the relationship between knowledge and ethics. It was in part this interest that led me to pursue medicine as a career path. As I progressed through medical school (University of Bergen, 2007–2013) I became increasingly fascinated by biomedicine and stories of clinical progress resulting from uncovering "truths" about the molecular origin of disease. One particular breakthrough that captivated me greatly was the development of imatinib as a treatment for chronic mye-loid leukaemia. Upon graduation, I therefore began training as a medical scientist as a PhD candidate in the Precision Oncology Research group at the University of Bergen (2013–2019). Through my training and work as a medical scientist, I became familiar with biomedical methodology and approaches, novel medical technologies, and emerging medical prac-tices. In my own project, I studied the variability and temporal dynamics of acute myeloid blood cancer; how cancer cells are similar and different, how cancer cells change in response to therapeutic interventions, time, and their environment, and how cancer cells reciprocally influence their surroundings. Studying this complexity and how this complexity related to the imaginaries of precision oncology I gradually became aware of what appeared to me as a significant discrepancy between the promises and expectations of biomedicine and the practical realisation of these promises. I also became increasingly intrigued by the rapidly expand-ing boundaries of medicine, science, and technology. In 2015 I attended the *"Ethical, Economical, and Societal Aspects of Cancer and Cancer Research"* course. This resulted to a radical shift in my approach to and understanding of precision medicine. I experienced that this course pro-vided me with perspectives, tools, and social resources which allowed me to think of and respond meaningfully to the tensions I was experienc-ing within the knowledge culture I was embedded. Following this course, I established a collaboration with Professor Roger Strand and Dr. Anne Bremer at the Centre for Study of the Sciences and Humanities (SVT), and I came to learn more about the complex processes governing the emergence of medical knowledge and essentially directing clinical care. As I developed these transdisciplinary collaborations, the dissonance I observed in precision medicine became ever more important and inter-esting to me. Learning of how precision medicine was not only a matter of science but also of ethics, sociology, and philosophy I gradually inte-grated language, thinking, as well as methodological approaches of theory of science, medical ethics, as well as science and technology studies in my own work. This led me to explore the potential and limitations of the precision medicine-related approaches I was concerned with through my own biomedical research in several dimensions. As part of my doctoral thesis "Exploring the boundaries of precision haemato-oncology - The case of FLT3 length mutated acute myeloid leukemia," I questioned the biological assumptions and theories underpinning precision medicine-related approaches in acute myeloid leukaemia, I explored precision medicine through ethical and social lenses, and I questioned the ultimate goals and purposes of medicine at large and of precision medicine in particular. Upon finishing my PhD, these interests led me to move towards places and spaces of medicine in the outskirts of technoscientific biomedicine. I am currently pursuing a specialisation in psychiatry while in parallel working as a postdoctoral fellow at SVT where I get to take engage in precision medicine as a practitioner of RRI, exploring more theoretically and philosophically questions related to the relationship between the normative and the scientific ways of knowing in medicine, and the relationship between human experiences and biomedical practices and imaginaries. (*Caroline Engen, University of Bergen*)

RRI in TRANSFORM, DLN, and CCBIO: similarities and contrasts

DLN and CCBIO were typical technoscientific endeavours. They were centres with ample funding for cutting-edge research in biotechnology, biomedicine, bioinformatics, and other natural sciences. RRI was introduced into them as mandatory and voluntary elements, respectively, with an explicit theoretical foundation in the AREA framework. From a distance, they could be seen as exemplars of how to "implement RRI" as soft, midstream-modulating governance of technoscience, mandated from the top of the organisation and with considerable resources in terms of staff and budget.

TRANSFORM was not a typical technoscientific endeavour. The Lombardy cluster did not interact with innovation processes or actors as such. In Catalonia and the Brussels-Capital region, some hands-on innovators took part in the work but not in central roles. Rather, the major roles belonged to public officials and citizens, as in Lombardy. Moreover, the conceptual framework for RRI was that of the European Commission keys as well as the methodologies of citizen science, participatory agenda-setting, and design thinking, all of which were wholly absent in DLN and CCBIO. When viewed from a distance, the two experiences could hardly be more different.

Staying at a distance and asking about "implementation" and "RRI impact," the similarities between TRANSFORM, DLN, and CCBIO are more striking. None of the endeavours appeared to achieve much visible and concrete governance of technoscience towards responsibility in the RRI sense, at least not if technoscience was imagined in the iconic forms, with white coats and machines and the cover page of *Science*. There were hardly any signs of midstream redirection or modulation of technoscientific research. In TRANSFORM, this was not seen because these technoscientific icons were not in direct view. In DLN, technoscience was definitely there, but there is little evidence that the scientists and technologists had any desire or intent to be governed and changed, and even less made any effort in that direction. And as we write that very sentence, we are struck by the counter-intuitive character of what was asked of them in the first place.

Furthermore, there is little evidence that there were concrete and direct efforts to govern them. With the exceptions we mentioned, this picture also holds for CCBIO. This is not to say that there were no visions for such changes. The problem of projectification that was explained in Chapter 3, however, meant that there was very little space for projects to be responsive in the RRI/AREA sense even if they wanted to. Especially in DLN, the projects were defined by proposals and grant contracts that described expected results and deliveries in detail. This was not compatible with open processes of agenda-setting for the research, and this may in turn explain why there were so few instances of real public engagement as well. In CCBIO, there was in principle more flexibility on the contractual level since the core grant was to a centre of excellence. However, in practice, the principal investigators operated in highly competitive research fields in which the competition for resources and academic capital severely constrained their degrees of freedom. Competitive research is as such a Red Queen situation where one cannot risk wasting time on thinking too far outside the box.

As a result, RRI integration efforts in DLN and CCBIO often came to materialise at a different temporal scale, or if not materialise, at least be imagined as something for the future. RRI teaching was definitely an instance of that: If principal investigators were not too interested in RRI, anticipation, reflexivity, etc., one could at least teach it to students who had to attend in order to get credits and who were easier to engage and convince. With this fact, a narrative emerged among the RRI staff and RRI persons that the students and the early career persons are the future and it is they who are going to change science so that it listens when society speaks back to it. Perhaps there is a grain of truth in this narrative; it is difficult to assess. As such the situation was not so different from that in TRANSFORM, which worked on other actors than the white coats, actors surrounding the technoscientists proper in the R&I ecosystem. In this way, TRANSFORM also acted on a different temporal scale if it was seen as governance of technoscience. The long temporal scale was even explicit in part of the RRI research taking place, as witnessed in the concluding paragraph of the second CCBIO anthology, after its final argument about how to incorporate a broader set of human values into cancer biomarker research:

> Such ideas may well not be easy to operationalise and implement. Indeed, we promised that this book will not make cancer biomarker research more streamlined or efficient in the short run. But we have not written this book for its thoughts to be implemented into practice in 2022, or even 2030. We have written it for 2100 and 2200. And while the old is dying, the new, in order to be born, has to be conceived and gestated in order to nourish and develop into existence.
>
> (Strand and Bremer 2022, 276)

The pattern from these three RRI experiences is that if the spatial distance to technoscience can be overcome, the temporal one is invoked with the rediscovery of the *longue durée* perspective.

Leaving the imaginations and looking closely at how RRI in fact was translated in the three experiences, the similarities are once again striking. There is capacity building, and there is maintenance and care work in relationships between human beings. In some cases, these relationships have a long history. In other cases, new relationships have to be formed, and care has to be invested in order to grow trust and overcome tensions and distrust. Is this how RRI is translated or even how it is best to translate it – into logics of care and interpersonal relations? Is this governance of technoscience? Our empirical studies all lead to these questions that will be discussed in more detail in our final chapter.

Notes

1 The personal experiences with DLN and CCBIO that are described in this chapter belong to one co-author, Roger Strand. In DLN, Strand was hired as adjunct professor to work on and with RRI. In CCBIO, Strand was one of the principal investigators and responsible for what at first was called the ELSA line of research, and later the Societal studies programme.
2 Readers with an affinity to George Orwell might be amused by the fact that the official Norwegian name of the Ministry of Education and Research is *Kunnskapsdepartementet*, which literally means "the Ministry of Knowledge."
3 For more information, see also the DLN website, https://www.digitallifenorway.org/. Accessed on April 7, 2023.
4 https://www.uib.no/en/svt. Accessed on April 7, 2023.
5 For more information about the CCBIO, see also its website, https://www.uib.no/en/ccbio. Accessed on April 7, 2023.

References

Åm, Heidrun. 2019. "Limits of Decentered Governance in Science-Society Policies." *Journal of Responsible Innovation* 6 (2): 163–78. https://doi.org/10.1080/23299460.2019.1605483.

Blanchard, Anne. 2016. "Mapping Ethical and Social Aspects of Cancer Biomarkers." *New Biotechnology* 33 (6): 763–72. https://doi.org/10.1016/j.nbt.2016.06.1458.

———, and Roger Strand eds. 2018. *Cancer Biomarkers: Ethics, Economics and Society*, 2nd edition. Megaloceros Press.

———, and Elisabeth Wik. 2018. "What Is a Good (Enough) Biomarker?" In *Cancer Biomarkers: Ethics, Economics and Society*, edited by Roger Strand and Anne Blanchard, 7–24. Bergen: Megaloceros Press.

Borch, Anita, and Harald Throne-Holst. 2021 "Does Proof of Concept Trump All? RRI Dilemmas in Research Practices." *Science and Engineering Ethics* 27 (7): 1–21. https://doi.org/10.1007/s11948-021-00335-4.

Bremer, Anne, and Roger Strand, eds. 2022. *Precision Oncology and Cancer Biomarkers: Issues at Stake and Matters of Concern.* Springer. https://doi.org/10.1007/978-3-030-92612-0.

Bremer, Anne, Elisabeth Wik, and Lars A. Akslen. 2022. "HER2 Revisited: Reflections on the Future of Cancer Biomarker Research." In *Precision Oncology and Cancer Biomarkers: Issues at Stake and Matters of Concern*, edited by Roger Strand and Anne Bremer, 97–119. Dordrecht: Springer.

Delgado, Ana. 2013. "RSB Policy Report: On Integrated ELSA Methods." *Zenodo.* https://doi.org/10.5281/zenodo.7870999.

_____, and Heidrun Åm. 2018. "Experiments in Interdisciplinarity: Responsible Research and Innovation and the Public Good." *Plos Biology* 16 (3): 1–8. https://doi.org/10.1371/journal.pbio.2003921.

Engen, Caroline Benedicte Nitter. 2020. "Exploring the Boundaries of Precision Haemato-oncology: The Case of FLT3 Length Mutated Acute Myeloid Leukaemia." PhD diss., University of Bergen.

Fisher, Erik, Roop L. Mahajan, and Carl Mitcham. 2006 "Midstream Modulation of Technology: Governance From Within." *Bulletin of Science, Technology & Society* 26 (6): 485–96. https://doi.org/10.1177/0270467606295402.

Gissum, Karen Rosnes, Sigrunn Drageset, Liv Cecilie Vestrheim Thomsen, Line Bjørge, and Roger Strand. 2022. "Living with Ovarian Cancer: Transitions Lost in Translation." *Cancer Care Research Online* 2 (4): e032. https://doi.org/10.1097/cr9.0000000000000032.

Gulbrandsen, Elisabeth. 2022. "Lange Horisonter og nye Fortellinger i Forskningspolitikken." *Nytt Norsk Tidsskrift* 39: 85–94. https://doi.org/10.18261/nnt.39.1.10

Hellström, Tomas. 2018. "Centres of Excellence and Capacity Building: From Strategy to Impact." *Science and Public Policy* 45 (4): 543–52. https://doi.org/10.1093/scipol/scx082.

Hesjedal, Maria Bårdsen. 2022. "Socializing Scientists into Interdisciplinarity by Placemaking in a Multi-Sited Research Center." *Science Technology and Human Values.* https://doi.org/10.1177/01622439221100867.

_____, and Heidrun Åm. 2022. "Making Sense of Transdisciplinarity: Interpreting Science Policy in a Biotechnology Centre." *Science and Public Policy* 50 (2): 219–29. https://doi.org/10.1093/scipol/scac055.

_____, Knut H. Sørensen, and Roger Strand. 2020 "Transforming Scientists' Understanding of Science–Society Relations. Stimulating Double-Loop Learning When Teaching RRI." *Science and Engineering Ethics* 26: 1633–53. https://doi.org/10.1007/s11948-020-00208-2.

Hesjedal, Maria B., and Roger Strand. 2021. "Transdisciplinarity in Digital Life Norway." White paper. Centre for Digital Life Norway. https://www.digitallifenorway.org/news/transdisciplinarity-in-digital-life-norway.html.

Nydal, Rune. 2006. Rethinking the Topoi of Normativity: Co-production as an Alternative to Epistemologically Modelled Philosophies of Science. Phd diss., Norwegian University of Science and Technology.

RCN (The Research Council of Norway). 2014. "Digital Life – Convergence for Innovation." White paper. Oslo: Research Council of Norway.

Solbu, Gisle. 2021. "Frictions in the Bioeconomy? A Case Study of Policy Translations and Innovation Practices." *Science and Public Policy* 48 (6): 911–20. https://doi.org/10.1093/scipol/scab068.

Strand, Roger. 2019. "Striving for reflexive science." *Journal for Research and Technology Policy Evaluation* 48, 56–61. https://doi.org/10.22163/fteval.2019.368.

Strand, Roger, and Lars A. Akslen. 2017. "Hva er Ansvarlig Kreftforskning?" *Tidsskrift for Den norske legeforening* 137: 292–4. https://doi.org/10.4045/tidsskr. 16.0295.

Strand, Roger, and Anne Bremer. 2022. "Conclusions: The Biomarkers that Could Be Born." In *Precision Oncology and Cancer Biomarkers: Issues at Stake and Matters of Concern*, edited by Roger Strand and Anne Bremer, 269–76. Dordrecht: Springer.

9 Governance in technoscience

Governance *of* or governance *in*?

We embarked on this endeavour of understanding various regional practices of (or attempts at) responsible governance of technoscience on a positive note: By looking into the abyss. We took de Sousa Santos' notion of "abyssal thinking" – the creation of otherness – as our point of departure to tell a story about technoscience. In this story, we travelled through Polanyi's (1962) seemingly self-governing Republic of Science, Kuhn's (1962) land of paradigms, and stumbled past Bush's endless frontier. We encountered a melancholic Leo Szilard, the nihilistic Marvin Minsky, and even got a glimpse of Victor Frankenstein and his wretched creation. Our further travels took us to the European Commission's Directorate of Research and Innovation in Brussels, where we learned about the keys to innovation governance, to the United Kingdom with its very own nautical history and ideas of responsibility. Finally, we ventured into Lombardy, to Catalonia, and then back to Brussels to meet people struggling to innovate responsibly, or if not, helping others to innovate responsibly. As the protagonists of our own story, we pushed forward, ever forward. But where to? And at what cost?

Our point of departure was the philosophical and historical emergence of a desire to govern technoscience. The desire to give a direction to the journeys of technoscience – to find "stewardship" in the words of Jack Stilgoe and his colleagues (2013) – that is more responsible or at least less reckless. This desire, we argued, challenges abyssal thinking and tries to find ways to move beyond deeply entrenched distinctions and imagine differently what it means to live in a world so fundamentally interwoven with the cultures and practices of technoscience (Appadurai 1996; Taylor 2002; Baptista 2014; Jasanoff and Kim 2015). In this book, we tell stories about a set of actors pursuing this desire. Furthermore, by following these actors and writing the book, we have tried ourselves to contribute to its enactment, by creating stories that perhaps facilitate different imaginations:

> [T]he imagination has become an organized field of social practices, a form of work (in the sense of both labor and culturally organized practice),

DOI: 10.4324/9781003371229-9

and a form of negotiation between sites of agency (individuals) and glob-
ally defined fields of possibility.

(Appadurai 1996, 31)

It is not at all trivial to make the next step and to actually imagine practices that would enact a Republic of Technoscience that moves beyond abyssal
thinking. What have we learned about translations of RRI and practices of
governing the Republic of Technoscience?

When examining how RRI is translated in the TRANSFORM project, we
do not primarily see activities that implement RRI principles into the regional
innovation context. What we observe instead is a variety of ways of engaging with actors and stakeholders in the regional innovation ecosystems that
are characterised by, and sometimes conditioned by, existing networks, procedures, shared concepts, and ways of working, specific for each regional
context. This general observation is difficult to reconcile with the idea that
responsible research and innovation (RRI) is about mainstreaming ready-made
tools, toolkits, and procedures that can be implemented with little regard for
geographical, historical, cultural, and political peculiarities. It led us to look for
alternative models of interpretation that could make sense of how the regional
context influences, and in some cases determines, what engagement practices
are chosen. We found that contrasting what we refer to as a *logic of implementation*, with a *logic of maintenance*, (inspired by Annemarie Mol's (2008)
contrast between the logic of choice and the logic of care, and Vinsel and Russell's (2020) work on maintenance) brought out this tension in the way RRI is
understood and put to work in innovation contexts. By "logic of implementation," we refer to the idea that across-the-board principles and procedures
developed in previous engagement projects can simply be applied in regional
innovation contexts and in that way foster enduring change in the way innovation is governed. In contrast to the logic of implementation, a "logic of maintenance" refers to activities, principles, and procedures that precede the actual
project, and to how these precedents are maintained and adapted throughout
the project life span, and finally how the project, after being "completed" in
accordance with the logic of implementation, enters a post-project afterlife,
where maintaining networks and social orders crucial for upholding the long
term relationships that make a continued focus on RRI possible can take place.
In this perspective, the project is *always already a predecessor-to-be*. From the
point of view of a logic of maintenance then, the project is a link in a chain of
activities where piecemeal tinkering and adaptation foster responsiveness beyond what the project-focused logic of implementation can capture. The logic
of maintenance may be negatively defined as that which is rendered invisible
by a logic of implementation. The logic of implementation is related to what is
to be fulfilled, completed, and generate impact. Maintenance is related to what
is open-ended, durable, persevering, and long term.

Thinking in terms of care and maintenance implies a distinct model of governance. This model does not aim for controlling, eradicating, or abandoning our Frankensteins. It stays with them (Halpern et al. 2016), "stays with the trouble" in Donna Haraway's terms (Haraway 2016).

When focus shifts from the logic of implementation to logic of maintenance, RRI ceases to be a reservoir of techniques available for application in diverse contexts. It becomes instead a way of thinking about responsiveness that is adaptive, receptive to the specific environments, or, in our preferred term, *translatable* to the specific needs of (regional) actors.

This shift towards care and maintenance resonates with Arie Rip's (2006) distinction between governance *of* and governance *in* – in his case, of and in reflexive modernity. Governance *of* something takes the modernist stance of positioning the actor outside and above the object of governance. This is visible in many of our modern heroes such as the captain, the pilot, the scientific genius, or even Victor Frankenstein. Rip generously describes this as almost a necessity of the process of governing and policymaking more broadly. A certain challenge is described together with its solution or at least with an attainable future state. This response is then to be carried out by the political actors in charge, who are accountable, and indeed must be separated from the actors themselves for the accountability to work. However, there are crucial shortcomings inherent in this classic control and command model:

> The illusion of the modernist actor is to just go for agency (that is, making a difference), and fail or be successful, not because of the strength of his agency, but depending on circumstances out of his control.
>
> (Rip 2006, 94)

Governance *in* the Republic of Technoscience then means abandoning this illusion of control and steerability and focus on what Rip refers to as "circumstances" and patterns of unintended and even unexpected effects. These circumstances are the conditions that make certain strategies or governance approaches succeed. It is not the "intrinsic characteristics" (Ibid, 89) of a certain policy intervention, but the "repair work" (Ibid) that happens in other places.

Throughout the empirical sections of this book we talked about such circumstances as the conditions and the broader research and innovation (R&I) ecosystems in the different regions while also pointing to many instances of maintenance practices that appeared to be crucial for the success of the cluster activities. Within a logic of implementation, these practices may be invisible or, if visible, marginalised as non-essential or misrepresented as "networking," "capacity-building," or similar shallow, instrumentalist categories. For example, months of project work in TRANSFORM were spent on making phone calls, writing e-mails, and having Zoom meetings where the actors exchanged stories and interpretations and aligned imaginations and expectations. Above all, the work of maintenance and care is hermeneutic and moral work. At the same time, it does not only sustain but also creates something new when for instance the endometriosis patients, doctors, researchers, and public administrators (of Chapter 5) come together, and their horizons of understanding come into contact. In the RRI mantras of Horizon 2020, the structure of such meetings was conceived to be of a particular type: Society was thought to "speak back" to Science, and what it was going to speak back about, was the needs and concerns of the citizens. What we show in our empirical chapters, is that needs and concerns sit at all ends, also in the need for maintenance of and care for the involved research-funding and research-performing organisations. In some of our examples, RRI becomes a matter of caring for public administrations, or at least caring for certain actors within the administrations who play key roles in upholding organisational commitments to certain lines of work, activities, and values. In this way, projects and initiatives such as TRANSFORM may support such actors "who do good." Here we may highlight another feature of the logics of care and maintenance and the idea of governance *in* and not *of* the system: These approaches are not afraid of playing the role of support and of not being the hero who controls, fixes, and rescues. It may remind us of Rip's (2006) idea of *reflexive* governance where the actor is conscious that her actions take place within the immanent plane of the other actors. We write "her" not to exclude men from reflexive governance but to emphasise the kinship to feminist critiques of modernity and feminist care ethics and also to remind us of the fact that all cluster leaders and almost all key personnel in the cluster leader organisations in the TRANSFORM project in fact happened to be women. The fact was remarked on several occasions, and we leave it to the reader to contemplate its significance.

Governing *what* now, exactly?

Throughout this book we have directed attention to the incredible diversity of practices that may take on the label of RRI. The question then becomes what exactly was the red thread, the mutual element of these practices that we have followed, apart from the prescriptions of the Horizon 2020 Science with and for Society funding stream? One perhaps cynical answer to this question would be that apart from trying to adhere to a particular call text there was no binding element and that the actors in the regions were doing what they would have done, or would have liked to be doing, anyway. If so, the main provision of the Horizon 2020 grant was simply its budget, that is money for salaries and activities that otherwise might have been difficult to pay for. Some statements of our colleagues could even be read as supporting this hypothesis, especially when they said that the actual label RRI did not necessarily mean that much to them.

We are not ready to settle for that cynical interpretation. First of all, it would contradict the preceding section, in which we postulated that the desire to govern technoscience responsibly was translated into the principle of RRI which again was translated as a myriad of practices that took the shape as maintenance and care work as a form of modest, immanent governance "in technoscience." It may rightly be asked, however, if the form was not too modest. Do these practices that we have described at such length, at all matter? Do they represent any deviation from business-as-usual? And last but not least, where is the technoscience to be seen in these stories? Indeed, we devoted a whole Chapter 8 to contrast the TRANSFORM story with stories about RRI initiatives in contexts where there were hi-tech laboratories. In TRANSFORM, not a single white coat was to be seen.

We shall return to the final question about the presence or absence of technoscience and begin with the middle one, about business as usual. In fact, it is not unreasonable at all to point out that the self-governance described by Polanyi as constitutive of the Republic of Science actually takes the form of governance *in* the Republic. In Polanyi's vision as well as in the entire tradition of modernist philosophy and sociology of science, from the logical positivists to Popper and Merton, the scientific community should never be ruled from above by a sovereign, a government, or other external actor. It should be self-governed in and by its own body politic (properly demarcated) and moreover by the universal, eternal, and self-evident norms dictated by the demands of curiosity, objectivity, and truthfulness. There is governance but only in the plane of action, keeping the ship in good order but never trying to take the steering wheel. It does not have a wheel or even a rudder. RRI, as we have seen it both in TRANSFORM and in the contrasting examples of Chapter 8, could be interpreted as yet another support mechanism that at its best helps keep the ship in good order and nothing more. That is, TRANSFORM did not interact much with the ship itself but rather with the dry dock, the shipyard, and the shipping company. From this perspective, things such as RRI do not make any sense, other than possibly building public and political support for science. As discussed in previous chapters, this perspective has never been absent, and definitely not in the higher echelons of political institutions in modern societies.

However, it will be obvious to the reader by now that the authors of this book do not share Polanyi's theoretical perspective; we consider it empirically refuted, theoretically flawed, and obsolete. Science, technology, and society are co-produced, and scientific practice is not dictated by universal and eternal norms. The Republic of Science is a Republic of Technoscience if not a Republic of Sociotechnoscience, and if we stay with the ship metaphor, it is probably more an armada than a single ship, and they do have rudders.

This may all sound as an argument leading to the conclusion that RRI is part of business-as-usual, maintains it, and supports it. RRI workers are also part of the crew on the ships of sociotechnoscience and members of its body politic. Is it all the same? As we asked above, does RRI matter?

In the actual empirical cases that we have been describing in this book, we find it easier and more plausible to argue that TRANSFORM mattered than the RRI efforts in the typical technoscientific contexts, that is, the Centre for Digital Life Norway and the Centre for Cancer Biomarkers. The similarity of all efforts is that they work through *maintenance, support and care.* Now, in TRANS-FORM one could support other actors who held a change agenda (such as the partners in the Catalan Generalitat), or who were willing to explore ideas for a change agenda (such as the Lombardy Region and the INNOVIRIS funding agency), or who experienced a neglected need (such as the endometriosis patients or the Brussels citizens in want of food supply). One could support those *"who do good"* in the very specific sense of trying to change the abyssal thinking, or those who had been made invisible by the same abyssal thinking. This is how sustaining something also can lead to change: One sustains and supports the potential for change. And we may witness governance in the armada as one of the ships somewhat changes course, which may affect the course of the entire armada. At least one may hope so.

The ships of Digital Life biotechnology and cancer research, however, had their courses plotted already in the non-transformational direction. In policy documents and frontstage policy discourse, the Centre for Digital Life Norway was presented and promoted as transformational, transdisciplinary, and RRI; however, this discourse did not correspond to the intentions, agendas, capacities, or practices to its actors, except the little minority who were assigned with RRI work. With the rudder set and no steering wheel in sight – and no access to it, had it existed – RRI work is hence left to do things that indeed felt meaningful but that hardly mattered in the sense of changing the abyssal course. One could liken it to rearranging the deck chairs on the Titanic, or with a less cruel comparison, to playing classical music while the ship moved towards the abyss, which arguably was a dignified and honourable act.

Rip (2006) rightly ironised over the modernist actor who "just goes for agency." A much-celebrated definition of governance was provided by the Carlsson report:

> the sum of the many ways individuals and institutions, public and private, manage their common affairs. It is a continuing process through which conflicting or diverse interests may be accommodated and cooperative action may be taken. It includes formal institutions and regimes empowered to enforce compliance, as well as informal arrangements that people and institutions either have agreed to or perceive to be in their interest.
> (Commission on Global Governance 1995)

Meaning-making contributes in this sense to governance in ways that our ship metaphor is not well suited to display. It is not the Centre for Digital Life as such that sinks, it is the armada of the modern sociotechnoscience that (according to our narrative) is on a destructive course. Therefore, it is not pointless to play classical music or engage in other dignified and honourable practices in the salon of the centre. Perhaps the "impact," to use that word, of RRI efforts, will take form in new thoughts within the minds of the young biotech researchers just as much as in the minds of the public administrators in Brussels, Lombardy, and Catalonia who already knew the need for transformation. Perhaps we should take it seriously that RRI is first and foremost a *principle*, that is, an idea. Strand and Spaapen (2020) described how the uptake of RRI in Horizon 2020 could be seen as a ship that sank, but continued that it lives on, in the shape of a "Flying Dutchman":

> […] that is, a ghost that might haunt, fascinate and inspire the minds of the thousands of sailors and fellow travellers who together constitute the collective governance of science of the future."
>
> (Strand and Spaapen 2020, 56)

This takes us back to the abyss and shows the potential of RRI being a spectre haunting actors while they do what they would have done anyhow. The spectre challenges deeply entrenched collective imaginations about what the legitimate role of innovation governance should be, how governance processes should look, and who should have a say in them. In Chapters 4–7 there were several examples that could be interpreted this way, especially on the level of public administration.

Indian mythological and philosophical thought has for millennia operated with three key types of agency: Creation, sustenance, and destruction. For life to flourish, all three must be present, in balance with each other. The European Age of Exploration was unbalanced. It wanted creation and sustenance at home and exported destruction to the colonies. Since then, the value of creation has gained importance to the extent that novelty has become a good in itself, and official, serious governmental policies have come to endorse concepts such as "disruptive innovation." If we think of RRI as a spectre, we can see it as reminding its experiencers of the value of sustenance and the need of a balance between the three types of agency. Consider, for example, that in the Horizon 2020 call text for which the TRANSFORM project was designed, applicants were asked to detail "*Specific, Measurable, Achievable, Realistic, and Time-bound*" impacts of their project proposals. The list abbreviates to "SMART," giving the set of criteria a strong positive value. The required impact details do not exclude maintenance but tend to steer the focus towards other aspects of RRI, towards the desirability and appreciation of project-induced, measurable changes in the engagement ecology, while relegating to the background, outsmarting so to speak, what upholds, stabilises, and safeguards the social orders that are necessary to *sustain* change. There is perhaps a risk that the strong emphasis on *change* in the impact evaluation of projects obscures the larger picture of how the expected impacts are absorbed by the innovation system and translated into long-term effects that consolidate social orders.

Maintenance thus proves to be a useful concept, precisely because it can serve as a contrast to a sometimes-excessive emphasis on what is new and changing, with the adjacent normative assumption that the change is desirable, and indeed SMART. Maintenance on the other hand is associated with continuity, preservation, durability, and requires time and continued attention. This is true for objects in the physical world, such as a bridge, a freshwater source, or a bike chain, but not less for relationships in the social sphere.

Does the logic of implementation serve to foster RRI, understood as "a transparent, interactive process by which societal actors and innovators become mutually responsive to each other with a view to the (ethical) acceptability, sustainability and societal desirability of the innovation process and its marketable products?" (von Schomberg 2011). Project-driven research governed by a logic of implementation promotes a focused and concentrated effort to create impact by addressing specific challenges in an effective way inside a defined time frame. On the other hand, the logic of implementation strongly favours projects that aim for specific, measurable, achievable, realistic, and time-bound impacts. What we see in the TRANSFORM project is that, paradoxically, the project seems to work most effectively and efficiently against a background of long-term, general, qualitative, and open-ended maintenance of social relationships, which is external to project ballistics, and invisible through the lens of implementation. And perhaps even more paradoxically to minds who have been influenced by contemporary research policies, it seems that the RRI principle is more effective when it reminds people of what they already knew or suspected somewhere at the back of their minds.

Preliminary conclusion: a tension in the presuppositions of RRI

The idea of RRI can be seen to build on two basic assumptions. Firstly, RRI only gives meaning if R&I are seen as social activities subject to normative valuations and oriented towards societal goals, ideally the common good. At the same time, the idea of *implementing* RRI in R&I processes presupposes that there is a lacuna in R&I that needs to be filled by a healthy dose of RRI. Because normative considerations of societal goals and orientations towards the common good are *different* from and even alien to the factually oriented and value-neutral R&I process, it is seen as an add-on. The first assumption is integrative, the second differential, and they work in opposite directions.

As we have alluded to, the demarcation line between R&I and RRI can be seen as analogue to the one between facts and values, the descriptive and the normative. The differential assumption then, leads to the idea that the responsibility for ensuring implementation of RRI (the normative end of the dyad), should be relegated to a specialised body of expertise, fitted with the appropriate skills and competencies to secure considerations of "ethical acceptability, sustainability, and social desirability of the innovation process." This neat dichotomy seems to "exhaust the field of relevant reality" (de Sousa Santos 2007) but serves simultaneously to "make invisible the abyssal line on which [it is] grounded" (Ibid), that is, the social fabric on the basis of which public engagement, citizens assemblies, or citizen science (and science itself for that matter) become at all possible. The maintenance work that was found to be *essential* to the RRI-related activities in the clusters of the TRANSFORM-project, belongs to what is lost in "the abyss," when implementation of RRI is equated with the value-dimension of R&I. With de Sousa Santos (2007) one could speak of maintenance in this sense, as falling victim to the abyssal thinking implied in the logic of implementation.

What happened to the technoscience?

Nordmann (2007) warned against speculative ethics that discusses, assesses, and validates hypothetical futures and thereby "squanders the scarce and valuable resource of ethical concern" (Ibid, 31). Instead, he argued, the ethical attention should be directed towards actual research and actual technoscientific developments. Indeed, many projects, activities, and funding programmes of the ELSI/ELSA/RRI type have been engaging with issues connected to science-driven, sophisticated biotechnology, nanotechnology, information and communication technologies, and other emerging technologies. By using terms such as "science-driven" and "sophisticated," we do not intend a sharp distinction but rather connote to the type of technoscience that relies on expensive laboratories, recent scientific findings, heavy computation, etc. We think of things such as CRISPR, fullerenes, and artificial intelligence and not so much of yoghurt, soot, and abaci, that is all. When there is talk of the need for responsible governance of technoscience, the imagination is often directed to the former category, the "sophisticated" things and practices, rather than the yoghurt, that is, unless one wants to present a counterargument against claims about the exceptional need for regulating, deliberating upon, and governing the sophisticated technoscience.

Only in one single chapter of this book (Chapter 8) did we discuss entities that mainly do sophisticated technoscience, and even then, there was little or no discussion of new and exciting technoscientific objects – no CRISPR, no liposomes that transcend the blood/brain barrier, no invasive software or firmware programmes. Is this really a book about the governance of technoscience? Where is the technoscience?

Our reply will hardly be a surprise to the reader. We believe that technoscience is everywhere. Firstly, in a very literal sense, it is actually present in devices and infrastructures that already are so deeply integrated into the lifeworld that there is no sense of estrangement anymore to render them as "sophisticated" and therefore problematic and in need of governance. In the case of TRANSFORM, the entire project had to change all its plans for human interaction, the kick-off meeting of the project taking place the same week as the first COVID-19 lockdown in Europe. The meeting that was supposed to take place in Milan was cancelled and instead the consortium "met" by using an internet service called GoTo Meeting. Spending an entire day on the internet in this way felt very exotic and tiring. Within a few months, however, services such as this one, Zoom, Teams, and others became standard work routine, and whatever we had had of principled concerns were fading away. Innumerous grammatical mistakes in the manuscript presented in this book were found and corrected by artificial intelligence built into Microsoft Word. Also, in the pilots, there is technoscience, such as in the serious smartphone game that the youngsters of Mollet del Vallès helped develop to improve waste collection in their municipality. What is different from the typical paper of Nanoethics is that the RRI work being done in these cases does not focus its ethical and political attention towards issues being raised by novel technoscientific objects. In yet another sense, what we have at hand is governance *in* technoscience, not governance *of* technoscience.

Secondly, and more profoundly, we contend that it would be a mistake to see the "sophisticated" technoscientific objects and practices as isolated from the rest of the Republic of Sociotechnoscience, as isolated from the rest of the world. Without committing too firmly to actor-network theory, it is close to a commonplace that the world in which we live is richly connected, and that things, in particular man-made things, have a hybrid existence that contains natural as well as social and cultural dimensions. We have already indicated the relevant networks in our three geographical clusters; all of them involved organisations that fund research that would qualify as sophisticated technoscience. As such, the white coats and expensive laboratories were only one degree of separation away. Governance in these ecosystems is thus governance in technoscience also in this regard; it is governance at what Allenby and Sarewitz (2013) would call Level III and perhaps Level II technology, that is, the larger sociotechnical systems in which the machines and other artefacts are embedded and connected.

At this point one might be caught in pre-modern nostalgia and propose RRI as the Return to Reason (Toulmin 2001) from an overly differentiated conception of rationality as modernisation and differentiation. This is not what we wish to contend. Indeed, when we discussed the "RRI mafia" in Chapter 3, that is, the emergence of a new subsystem around RRI calls in Horizon 2020, we implicitly provided evidence against excessive nostalgic hopes in morality and wisdom through de-differentiation. Latour may have been right that we were never modern in the ontological sense but still there seems to be no way around the need for differentiated practices and bodies of knowledge in a technoscientific civilisation. We already indicated above, however, several reasons why the direct approach of trying to govern the white coats and their machines may not always be the most effective. If we are right that an important aspect of governance in technoscience is meaning-making and care work within a logic of maintenance, and that the desire to counteract abyssal thinking implies that certain values and intents should be maintained and cared for more than others, then the maintenance and care efforts should be directed towards sites and actors that embody and enact those values and intents. Often, this will be outside the sophisticated laboratories, as we have seen: In the municipality, at the hospital, in the public administration.

Epilogue

> The notion of 'steering', with its implication of an agent faced with an 'object' to be steered, is of course misleading since the steering agent is part of an evolving system including the 'object' and himself. To keep the tensions visible, I will use the term non-modern steering, a <u>contradictio in terminis</u>, as a programmatic concept.
>
> (Rip 2006)

The ship-of-state model, to which Rip alludes, and which Plato developed in the *Republic*, has followed us throughout the book. The art of good steersmanship belongs, according to Socrates, to the *manual* arts and ranks low in the hierarchy of knowledge, since it does not involve the use of reason (*logos*) and is not concerned with causes (Keyt 2006). Steersmanship consists instead in conjectures, and in the exercise of perceptions by practice and experience, "with the additional use of the powers of guessing" (Plato, *Philebus* 55e). The steersman is not someone who is "faced with an object to be steered" but is dependent on and adaptive to the characteristics of the vessel and its sailors, passengers, and cargo, to the shifting weather conditions, the ship's contractor, the shoreline and reefs, changing seasons, darkness and light, the capriciousness of the gods, and all the other things that pertain to the craft (*Republic* VI, 488). This steersman is part of evolving systems, and in this sense exemplifies the notion of governance *in* complexity, or indeed, "non-modern steering." Plato's allegory has been criticised by some for mistaking political power, which allegedly is concerned with the *aims* of political action, with the actions which will lead to the aims, the means. Thus, it is argued, a captain on a ship belongs to the latter category and is not a fitting metaphor for political power (cf. Bambrough 1956; Walzer 1983). The criticism seems anachronistic and in contradiction with the goal of steersmanship defined as "safety while sailing" (*sôtêria en tô plein*) (*Republic* I. 346a8). The primary function of steersmanship is not about projectile impact, or about a pre-defined destination, but about safeguarding the ship and everyone in it while at sea.

References

Allenby, Braden R., and Daniel Sarewitz. 2013. *The Techno-Human Condition*. Oxford: The MIT Press.

Appadurai, Arjun. 1996. *Modernity at Large: Cultural Dimensions of Globalization*. Minneapolis and London: University of Minnesota Press.

Bambrough, R. 1956. "Plato's Political Analogies." In *Philosophy, Politics and Society*, edited by P. Laslett, 105. Oxford: Macmillan.

Baptista, João Afonso. 2014. "The Ideology of Sustainability and the Globalization of a Future." *Time and Society* 23 (3): 358–79. https://doi.org/10.1177/0961463X11431651.

Commission on Global Governance. 1995. *Our Global Neighbourhood: The Report of the Commission on Global Governance*. Oxford: Oxford University Press. https:// unesdoc.unesco.org/ark:/48223/pf0000100074.

Halpern, Megan K., Jathan Sadowski, Joey Eschrich, Ed Finn, and David H. Guston. 2016. "Stitching Together Creativity and Responsibility: Interpreting Frankenstein Across Disciplines." *Bulletin of Science, Technology and Society* 36 (1): 49–57. https://doi.org/10.1177/0270467616646637.

Haraway, Donna. 2016. *Staying with the Trouble. Making Kin in the Chthulucene*. Durham: Duke University Press.

Jasanoff, Sheila, and Sang-Hyun Kim. 2015. *Dreamscapes of Modernity: Sociotechnical Imaginaries and the Fabrication of Power*. Chicago: University of Chicago Press.

Keyt, David. 2006. "Plato and the Ship of State," In: The Blackwell Guide to Plato's Republic, edited by G. Santas. Oxford: Blackwell Publishing Ltd.

Kuhn, Thomas S. 1962. *The Structure of Scientific Revolutions*. Chicago: University of Chicago Press.

Mol, Anemarie. 2008. *The Logic of Care: Health and the Problem of Patient Choice*. New York: Routledge.

Nordmann, Alfred. 2007. "If and Then: A Critique of Speculative Nanoethics." *Nano Ethics* 1 (1): 31–46. https://doi.org/10.1007/s11569-007-0007-6.

Plato. 1901[~350 BCE]. *"Philebus."* In Platonis Opera, ed. John Burnet, vol. II, 11–67. Oxford: Clarendon Press.

_____. 1902[~350 BCE]. *"Republic."* In Platonis Opera, ed. John Burnet, vol. IV, 327–621. Oxford: Clarendon Press.

Polanyi, Michael. 1962. "The Republic of Science: Its Political and Economic Theory." *Minerva* 1 (1): 54–73.

Rip, Arie. 2006. "A Co-Evolutionary Approach to Reflexive Governance – and Its Ironies." In *Reflexive Governance for Sustainable Development*, edited by Jan-Peter Voß, Dierk Bauknecht, and René Kemp, 82–102. Cheltenham/Northampton: Edward Elgar Publishing. https://doi.org/10.4337/9781847200266.00013.

Schomberg, René von. 2011. *Towards Responsible Research and Innovation in the Information and Communication Technologies and Security Technologies Fields*. Luxembourg: *Publication Office of the European Union*. https://doi.org/10.2777/58723.

Sousa Santos, Boaventura de. 2007. "Beyond Abyssal Thinking: From Global Lines to Ecologies of Knowledges." *Canadian Parliamentary Review* 30 (1): 45–89.

Stilgoe, Jack, Richard Owen, and Phil Macnaghten. 2013. "Developing a Framework for Responsible Innovation." *Research Policy* 42 (9): 1568–80.

Strand, Roger, and Jack Spaapen. 2020. "Locomotive Breath? Post Festum Reflections on the EC Expert Group on Policy Indicators for Responsible Research and Innovation." *Assessment of Responsible Innovation*, November, 42–59. https://doi.org/10.4324/9780429298998-4.

Taylor, Charles. 2002. "Modern Social Imaginaries." *Public Culture* 14 (1): 91–124.

Toulmin, Stephen. 2001. *Return to Reason*. Boston: Harvard University Press.

Vinsel, Lee, and Andrew L. Russell. 2020. *The Innovation Delusion : How Our Obsession with the New Has Disrupted the Work that Matters Most*. New York: Random House.

Walzer, Michael. 1983. *Spheres of Justice*. New York: Basic Books. 286.

Index

For Product Safety Concerns and Information please contact our EU
representative GPSR@taylorandfrancis.com
Taylor & Francis Verlag GmbH, Kaufingerstraße 24, 80331 München, Germany